Rebuildi

In the scholarship of urbanism, small towns are overlooked and understudied. *Rebuilding the American Town* highlights how smaller municipalities are transforming to serve their communities and meet the future. The book uncovers creative planning and design strategies of nine U.S. towns as they rebuild to remain vibrant, equitable and viable in the face of metropolitan sprawl, population shifts, political division, economic shortfalls and climate change. *Rebuilding the American Town* includes interviews and insights from those directly involved, to reveal the challenges and advantages of being a smaller city while highlighting the power of design at local levels.

The book provides a new lens for contemporary urbanism more broadly as it shifts thinking away from large-metro concerns, toward novel, tactical strategies that advance the quality of life for residents through design and policies that are scaled to the populations and places they serve. The projects in this book show how the small town in the United States is unexpectedly progressive, experimental, urban and global.

David Gamble is Lecturer in Urban Design at the Massachusetts Institute of Technology. He is a registered architect and certified urban planner and Principal of Gamble Associates, based in Cambridge, MA. The practice focuses on urban revitalization and community development. He is the co-author of *Rebuilding the American City* and editor of *Idea City*.

Patty Heyda is Professor of Urban Design and Architecture at Washington University in St. Louis. She researches American cities and design politics, with a focus on mapping and spatial justice. She is the author of *Radical Atlas of Ferguson, USA* and co-author of *Rebuilding the American City*.

"*Rebuilding the American Town* is a love letter to our small cities and towns and confirms that spectacular urban design and planning outcomes can be found in the most unlikely places. Towns deserve to be our true laboratories for urban innovation and regeneration. It is hard work and the process is not always tidy but this book shows us how it is being done. We need hundreds of such rebuilding initiatives across America!"

Maurice Cox, *Professor, Former Commissioner of the Chicago Department of Planning and Development and Detroit Department of Planning, and Former Mayor of Charlottesville, VA*

"This book brings new attention to the American town in a way that fundamentally reshapes how we think about community and progress in the 21st century. The case studies offer a template for how we refashion American towns into models of affordability, upward mobility, and resilience in the decades to come. A must-read for anyone who dreams of how to shape our civic life."

Shaun Donovan, *U.S. Secretary of Housing and Urban Development (2009–2014); CEO and President, Enterprise Community Partners*

"The well-documented case studies in this timely book will be a useful resource for years to come to everyone interested in helping small towns leverage local conditions to address challenges ranging from justice and climate change to depopulation and gentrification. Diverse and innovative improvements are discussed from various viewpoints, including candid insider insights on how they were financed and how they have performed both culturally and economically."

Ellen Dunham-Jones, *Professor of Urban Design, Georgia Institute of Technology and co-author of* Retrofitting Suburbia

"Americans tend to sentimentalize small towns in memory while having largely abandoned them for the lures of the metropolis. Patty Heyda and David Gamble present a prescient counterpoint, revealing the urbanity, entrepreneurship, community and fulfilling lives found in modest sized settlements across the country. *Rebuilding the American Town* is inspired in scholarship and perspective. As life-work expectations change facilitated by accommodating technologies, the narrative unveils a promising future for the American small town."

Alex Krieger, *Professor Emeritus of Urban Planning and Design, Harvard University, and author of* City on a Hill: Urban Idealism in America from the Puritans to the Present

"Like its *Rebuilding the American City* predecessor on larger cities, this companion volume's excellent case studies are packed with extremely useful information and are written in a very readable nontechnical style for a wide audience. Particularly helpful are the numerous interviews with community leaders, businesspeople, investors, design professionals and others who compellingly describe past successes and continuing challenges. An inspiration to read."

Randall Arendt, *author of* Rural by Design

Rebuilding the American Town

Design and Strategy at Small Scale

David Gamble and Patty Heyda

NEW YORK AND LONDON

Designed cover image: © David Gamble and Patty Heyda

First published 2025
by Routledge
605 Third Avenue, New York, NY 10158

and by Routledge
4 Park Square, Milton Park, Abingdon, Oxon, OX14 4RN

Routledge is an imprint of the Taylor & Francis Group, an informa business

© 2025 David Gamble and Patty Heyda

The right of David Gamble and Patty Heyda to be identified as authors of this work has been asserted in accordance with sections 77 and 78 of the Copyright, Designs and Patents Act 1988.

All rights reserved. No part of this book may be reprinted or reproduced or utilised in any form or by any electronic, mechanical, or other means, now known or hereafter invented, including photocopying and recording, or in any information storage or retrieval system, without permission in writing from the publishers.

Trademark notice: Product or corporate names may be trademarks or registered trademarks, and are used only for identification and explanation without intent to infringe.

Library of Congress Cataloging-in-Publication Data
Names: Gamble, David, 1968- author. | Heyda, Patricia Lynn, 1972- author.
Title: Rebuilding the American town : design and strategy at small scale / David Gamble and Patty Heyda.
Description: New York, NY : Routledge, 2025. | Includes bibliographical references and index. | Identifiers: LCCN 2024028190 (print) | LCCN 2024028191 (ebook) | ISBN 9780367258528 (hardback) | ISBN 9780367258535 (paperback) | ISBN 9780429290206 (ebook)
Subjects: LCSH: City planning--United States. | Urban policy--United States. | Small cities--United States.
Classification: LCC HT167 .G347 2025 (print) | LCC HT167 (ebook) | DDC 307.1/4160973--dc23/eng/20240820
LC record available at https://lccn.loc.gov/2024028190
LC ebook record available at https://lccn.loc.gov/2024028191

ISBN: 978-0-367-25852-8 (hbk)
ISBN: 978-0-367-25853-5 (pbk)
ISBN: 978-0-429-29020-6 (ebk)

DOI: 10.4324/9780429290206

Typeset in Avenir
by SPi Technologies India Pvt Ltd (Straive)

To my mother Catherine Gamble, who taught me to love my small town.
(DG)

For my family, I love you.
(PH)

Contents

Acknowledgements — viii

Introduction: American Towns in the 21st Century — 1

Section 1 Constellations — **21**

1.1 North Adams, MA — 22

1.2 Jamestown, NY — 48

1.3 Bentonville, AR — 76

Section 2 Satellites — **99**

2.1 Caldwell, ID — 100

2.2 Piqua, OH — 126

Section 3 Metroposts — **149**

3.1 San Ysidro, CA — 150

3.2 Opa-locka, FL — 176

Section 4 Outposts — **203**

4.1 Traverse City, MI — 204

4.2 Greensburg, KS — 228

Index — 254

Acknowledgements

This book is a collaborative project that follows on our first co-authored book, *Rebuilding the American City* (2016). Similar to that endeavor, we found our interviews and conversations with people in the towns we wrote about the most rewarding part of the research. We are indebted to everyone who generously shared their time and insights with us, mostly in person, but also over zoom during the COVID-19 pandemic when travel wasn't possible. We strived to maintain accuracy and reflect the lessons learned in the edited perspectives section of the nine cases. We're inspired by the work that is being done across sectors, neighborhoods and capacities, both on the ground in self-determination and from various leadership and supporting positions.

There are two special people we wish to thank, without whom *Rebuilding the American Town* would not be as beautiful and comprehensive as it is: Kelly Lynema created the information graphics with us, following her work on the first book. Kelly, thank you for your clarity and brilliance with these diagrams every time! And for your enthusiasm and partnership over quite a few years now. And thank you Xudong Zhu, whose exquisite graphics provided the uniformity and legibility for the 2d base maps and 3d aerial perspectives. Xudong, we are indebted to your commitment and the talents you shared with us on this project. Both Kelly and Xudong were valued, inquisitive graduate students at Harvard University's Graduate School of Design (GSD) when this project started and they stuck with us well past graduation and into their own professional urban design and planning careers.

Many others helped us along the way. Thank you, Lisa Regan at TextFormations, for your patience and skill helping us carefully edit tens of thousands of words from the rich conversations we had with those involved in rebuilding towns across the country. We would also like to thank our editors at Routledge, Selena Hostetler, Kathryn Schell and Gail Welsh, who stuck with us through the pandemic and the long process of completing the work. Funding for editing was made possible by the Office of the Director of the College and Graduate School of Architecture, and Urban Design at the Sam Fox School of Design & Visual Arts, Washington University in St. Louis and research travel was made possible by real estate grants from the GSD. The Sam Fox School at Washington University provided invaluable graduate and undergraduate research assistant-support over the years. In addition to Heather Woofter in the director's office, thank you to the Sam Fox School Dean's office and Nicole Allen, Ellen Bailey, Audrey Treece, John Foughty and to Linda Samuels, Derek Hoeferlin and Monica Rivera.

We're incredibly proud of our students in Architecture and Urban Design at Washington University in St. Louis, at Harvard's GSD and in the Department of Urban Studies and Planning at the Massachusetts Institute of Technology (MIT) who provided insights and inspiration for this book through our architecture, urban design and planning classes over the years. Our students also supported us as valued research assistants. At Washington University, thank you Celine Haddad, Margaux Elliott, Catherine Hunley and Weicong Huang for jumping in and assisting at key times, helping with images and transcribing interviews. We are also grateful for the initial research efforts of Miriam Keller and in Boston, Elyse Cote, Lexi Usher, Erin Gleason, Philipp Maue, Brian Gregory and Stephanie Vojvodich who helped to transcribe interviews, update graphics or just provide perspectives on the dynamics associated with small town revitalization.

Thanks also to colleagues Rahul Mehrotra, Diane Davis, Eve Blau, Rick Peiser and Ann Forsyth at Harvard for organizing conversations around small towns (both domestic and global) which encouraged us to continue our research and to Jim Stockard for his insights into community development corporations. At MIT, thank you Jeff Levine, Brent Ryan and Andres Sevtsuk for sharing your thoughts about small town revitalization and municipal leadership. At Washington University in St. Louis, much appreciation goes to colleagues in the Master of Urban Design program and to Liam Otten, for helping us share our work. Thank you also to the inaugural Urban Design Academic Council steering committee for inter-institutional support and our conversations on all things urban design.

Thank you to Alex Krieger, for his support and keen eye on the project as it evolved and—together with Anne Mackin—for the informal dispatches of Main Streets from across the American south. Alex is a mentor and good friend of both of ours and someone from whom we never stop learning. Thanks also to our colleagues from across the professional development, non-profit and academic worlds who were willing to read drafts and provide precious feedback and insights: Lee Einsweiler, Barry Alberts, Kelly Kost, Brent Ryan, Peter Lombardi, Craig Seymour, Larry Cranor, Brent Crittenden and Kennedy Smith. Thanks also to Kirk Narburgh, Scott Compton and Bill Hercules at the American Institute of Architects who shared initial research on development dynamics in rural areas.

Colleagues and representatives in the respective towns have been invaluable. In Bentonville, Arkansas, thank you to Michael Repovich for the tours and local knowledge, and Elysse Newman for the historical perspective. In Opa-locka, Florida, thank you Tanya Williams at Ten North who stopped work in the middle of a September day to show us around and share her deep knowledge of the town and the CDC. Chris Schmiesing and Paul Oberdorfer in Piqua, Ohio have been relentless in advancing downtown's revitalization, together with entrepreneurs like Joe, Laurie and Wilson Reiser, Jason and Kristy Moore and Jeff and Emily Pedro. In Traverse City, thank you to Marsha Minervini for making many connections and to residents Bob and Nancy Giles at the Village at Grand Traverse Commons who graciously opened their historic unit (and for their charcuterie!). In Greensburg, we are grateful to City Administrator Stacy Barnes and the leadership and staff at Greensburg's Big Well Museum for their hospitality. In Maine, thank you to Antony Muench and Randall Arendt for calling our attention to the good things happening in Caldwell, Idaho and for host Jeff Root upon arrival to Boise.

There are many others who we have inadvertently left off this list, especially those we interviewed who provided insights into towns we weren't able to include in this volume. We are grateful for your time and input. Lastly, we thank our families who have been through this before, for their immense patience and support. On the west coast, thank you Harriet Heyda. On the east coast, thank you Karen Heyda Jackson and Ivana Heyda. And to Bret, Thomas, Jack and Bridget, your love and support means more than you know. Jack, your transcription was key! Thank you Melissa and Elle, for your enduring support.

Introduction

American Towns in the 21st Century

In the scholarship of urbanism, small towns are overlooked and understudied. The research in this book provides a new lens for contemporary urbanism studies: a lens that focuses on smaller metros to learn how they advance qualities of life through creative, tactical and spatial transformations commensurate with their populations. *Rebuilding Towns* follows our 2016 study of redevelopment paradigms and creative design strategy in fifteen major U.S. cities (Gamble and Heyda, 2016).

Rebuilding the American Town highlights creative urban design and planning strategies in nine U.S. towns of approximately 50,000 people or less.[1] We say *approximately* since over the course of researching and writing this book, two of the towns featured have grown so rapidly that they exceeded that threshold. Others are shrinking, while some remain stable. The nine geographies, nonetheless, make visible the planned and unplanned complexities and tools that American towns draw on to remain vibrant and viable in the face of metropolitan sprawl, climate change, pandemics, privatization, and increasingly polarized political and social environments. Despite daunting challenges, the towns in this book are unexpectedly progressive, experimental and urban.

Map overlay of nine towns (in orange) featured in this book, together with the 15 cities featured in *Rebuilding the American City* (Gamble and Heyda, 2016).

We wondered—how does being small impact the dynamics, challenges or possibilities of rebuilding and redevelopment at a local level? It's a question that intersects with a number of overlapping national themes, discussed below. It is also a question that allowed us to explore what exactly rebuilding means, and for whom.

National Themes Impacting *Why Towns Now*

The Rise of Remote Work

COVID-19 (and its ever-evolving variants) has uprooted lives and altered, perhaps forever, our life/work balance. As a result of the pandemic, virtual work (which has been around for decades but is more feasible now with high-speed internet and other advances in technology) is enabling job opportunities in areas previously considered too far removed from urban financial centers. Since the pandemic, people with means have sought other spaces and *found them*. Many escaped to smaller communities where they could buy more with less money, or have easier access to the outdoors while maintaining work from home (Fox, 2020). Simply put, virtual working has made making a living in a town—even a remote one—more viable. Distance, not density, appears to be an advantage.

Political and Social Crises

Political polarization, increased social unrest, and a reckoning of systemic racism and inequality have resulted in new forms of justice advocacy work. Yet as some national media outlets amplify a false narrative that large cities like Los Angeles, Chicago and New York City are crowded and violent, others ignore the rising inflation, loss of public support and the escalating cost of living in these places. Governmental austerity measures and an erosion of social services for populations that need them most have continued unabated under both Republican and Democratic administrations (Heyda, 2024). As of December 2023, the United States has witnessed a 12 percent increase in homelessness, reaching its highest level according to the U.S. Department of Housing and Urban Development.[2] While these disparities are pronounced in large U.S. metros, the small town has not been immune to the issues and their impacts.

Against the backdrop of municipal distress, there is an increasingly fictionalized nostalgia around small towns as places of relative homogeneity and security—family values and American patriotism—that has become weaponized to stoke racialized rifts between urban and rural; big cities and small towns; liberalism and conservatism (Kelly, 2023). These are rifts that, we argue, do not really exist. Instead, the cases in this volume show how urban actors in smaller municipalities operate as they cross categories, even political platforms, to reimagine their public realms and rethink their towns' most relevant futures.

A big gun in a small town at the Daisy Air Rifle Museum in Rogers, Arkansas.

Inflation and the Nation's Affordability Crisis

Since the 1970s, the U.S. government has shifted away from direct public affordable housing provision to more privatized models. If the real estate boom elevated municipal tax coffers in cities, it also amplified an housing affordability crisis that now drives the working and middle classes away from the largest urban centers, or into new shared residential arrangements. More permissive policies around remote work and short-term rentals, together with a rising wariness of close quarters, accelerate that trend (Means, 2020).

Nationally, the production of housing at any scale has not met demand, and the mismatch between current household sizes and housing stock is genuinely lopsided (Gamble, 2023).

Housing prices dictate to a large extent where people live. Fast-growing metro areas, even many secondary markets, have simply become too expensive as they become too spread out to sustain healthy lifestyles and a healthy planet. However, as planners and engineers reproduce urban models that are exclusive or auto-dependent with environmentally destructive commutes, towns offer economic and scalar advantages. For many, the American Dream of owning one's own home is more attainable in a town versus a major metropolitan area.

Climate Change and Environmental Impacts of Sprawl

In *Visualizing Density*, Julie Campoli and Alex MacLean describe conditions by which cities in the U.S. need to re-evaluate their suburban land use patterns:

> How long we can sustain ourselves on our finite land mass will depend on how carefully we use land in this century…. As we confront dwindling land and energy resources, a concentrated growth pattern makes more and more sense: reuse land that has already been altered; limit the range of new development to an area that is easily accessible; and build up, not out. It is becoming clearer that these ideas should not be relegated to our past, but are the key to our future.
> (Campoli and MacLean, 2007: 5–6)

Small towns, especially those built in the late 19th and early 20th century, offer many of these advantages; their blocks and core infrastructure were densely built commensurate to their populations, already sustainably concentrated for walking and biking.

At the same time, the impacts of global climate change have brought intense heat, flooding and sea-level rise, fires, drought events and more severe storms to cities and towns alike across North America. As we witnessed in Lahaina, Hawaii, after wildfires and in Fair Bluff, North Carolina after Hurricane Matthew flooding, these impacts are restructuring economies and geographies of metro and rural regions as they are forced to rebuild after devastation, or move away, as climate migration becomes a reality (Flavelle, 2021). Like during the COVID-19 pandemic, people are also leaving vulnerable places to work remotely in other cities and towns that are more resilient, or that offer a means to get outdoors where weather conditions are still reasonable or conducive. While the towns in this book are impacted by different dimensions of the climate crisis, they offer models for how strategic actions can be applied to mitigate environmental impacts more locally. At the same time, with fewer people, financial capacities are more constrained in a small town, and funding the necessary infrastructural projects remains a challenge for isolated locales. Moreover, since most climate mitigation projects are large-scale and systemic, they depend on wider support and inter-governmental collaboration.

Greensburg, Kansas after a 2007 tornado of unprecedented strength wiped out the entire town. With more extreme weather events from climate change, cases like this will increasingly challenge what we mean by *rebuilding* American towns.
Credit: KEYSTONE_AP/ Orlin Wagner.

The Aura of Main Street

The underlying power of a small town is at once social and emotional (Orvell, 2012). Many older downtowns are lined with well-crafted historical buildings, even as some remain hidden behind layers of neglect or inexpensive fixes. Nevertheless, towns often leverage their architectural heritage as an amenity to celebrate the heyday of an earlier place and time, and to attract new markets and users. Indeed, the image of a walkable, mixed-use downtown with ground-floor commerce and upper-floor residences is still very much a part of

The Main Street atmosphere of Old Town Alexandria, Virginia. Credit: Kennedy Smith.

the American psyche. With their compact form, walkability and historic backdrops, towns offer such an experience, even if only a portion of that character remains intact today.

Still, the economic realities that small municipalities face are formidable. Despite the allure of their historic fabric, local city governments and private actors struggle to maintain their inventory. For many towns, the job opportunities and income levels needed to occupy vacant space simply are not there (and may never be again). In weak market economies, historic main streets are typically densely built but only mid-rise, making profitable redevelopment a challenge. Conversely, in popular historic districts rising land values stymie smaller independent businesses from affording Main Street rents. Regardless of where they are in the spectrum, towns struggle to fill storefronts as economies and the populations that initially drove them have shifted. The cases in this book explore the various ways town leaders have reconsidered these conditions and challenges. We see, for example, Main Street strategies that shift from retail-centric approaches to more public space-oriented priorities that reclaim Main Street and the downtown area as a center of collective public life.

Population Loss

Small towns are not all growing. For younger people, towns lack not only the high wage jobs but also the diversity of more urbanized areas. Public transit is minimal, if it exists at all. Despite stereotypes of a town's strong "sense of community," towns paradoxically may also perpetuate a greater sense of individualism that undermines inclusion. The towns we investigated did not offer a single solution to the population shifts that are ongoing. Some, like Greensburg, Kansas, rebuilt state-of-the-art internet connectivity to draw remote workers. Traverse City, Michigan relies on its airport to bring a younger, more geographically fluid labor force. Others, like the hamlets in Upstate New

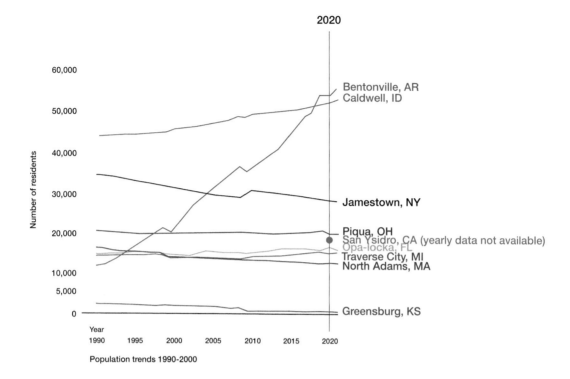

The nine U.S. towns in this book and their population trajectories since 1990.

York, have focused on tourism and recreation given their picturesque settings and proximity to major cities along the Northeast corridor. Since towns are small, they offer a test case for rethinking the urban growth paradigm altogether. To be sure, if cities have transformed according to too much outward urban growth—and unfettered capital growth—at the cost of increased disparity and the environment, then population stasis may not be a negative, but a necessity (Hickle, 2020). If towns can stabilize a balance of public services and access to jobs, they offer an important model of urbanization as *degrowth*.

Shared Trends Across Towns

Towns, like cities, are in a constant state of transformation. Leaders strategize how to recruit capital investment or tourism, as markets and industries transition. They strategize how to plan for climate change as weather events become more pronounced; and how to improve the character of the public realm for residents as governmental austerity measures continue apace. *Rebuilding the American Town* captures these dynamics at a critical inflection point in American urbanism, one where a number of useful insights have emerged.

Scale Matters

Planning in small towns is unique from that of larger cities in the ways political leaders interact with the population and in the political and economic models for financing projects. There is a very real profile of relationships that eliminates larger city dynamics of anonymity. Local decisions are personal; the smaller the town, the more elected leaders know their constituents on a first name basis. Accountability is harder to skirt. And while small towns compete with each other or with the larger metro region for basic fundamental infrastructure and resources, being in the shadow of the metro can also bring significant opportunities and clout. But not always. What works to attract investments and balance redevelopment in the American city and its surrounding region may not easily translate to the smaller nucleus. Nevertheless, we found that when alternate strategies were needed, they were often forged from circumstances and close relationships between people across sectors that facilitated implementation. In this way, many towns are achieving more contextual urban design and planning than larger cities.

In addition, while public–private partnerships are the paradigm of any American city of any size, towns like Columbus, Indiana (pop. 50,391)[3] reveal how the role of just one outsized company can have a transformational impact. In that case, the town has become a showcase of modern architecture and emerging design, funded by the Cummins Company in a relationship that dates back decades (Loos, 2017). In this book, Jamestown, New York and Bentonville, Arkansas demonstrate similar exponential impacts of their hometown companies and foundations.

Towns Bear Trauma and Provide Collective Support

Following years of systemic disenfranchisement and exploitation, underserved populations are increasingly living in American towns, large and small. This dynamic is not exclusive to historic majority Black towns like Kinloch, next door to Ferguson, Missouri, but also applies to rural settings with majority white populations. These towns experience collective trauma after years of hardship, a trauma which is impactful in smaller municipal footprints where public sector capacity is compromised. But as American Studies scholar George Lipsitz reminds us, in the face of segregation, there is meaningful congregation (Lipsitz, 2007). Systems of collective support, community and healing are enabled by processes of local rebuilding in Opa-locka, Florida and San Ysidro, California, the two towns featured in this book that have systematically been disinvested and contested. The role of Community Development Corporations (CDCs) in these towns is notably strategic and creative. In rural Greensburg, Kansas, an older and majority white demographic also foregrounded collective community healing throughout their process of rebuilding after a traumatic tornado literally erased the town. The temporary tent that residents and leaders met under for over a year brought people together by serving as both formal meeting hall and informal town square for interaction—and an important collective BBQ space.

Towns Are Global

Many agrarian towns have lost farming jobs due to changing global climate instability and extreme weather, or due to the growing scales of corporate agricultural (Big Ag) and industrial and fossil energy production. These factors sever connections between towns and the land that surrounds them, as some deal with the further injuries of depleted water and poor soil and air quality. Yet, these issues have led to unexpected new models of American town-making: The very conservative community in Greensburg rallied around "progressive" sustainability principles as they rebuilt "green" to reassert their connection to the land. In other cases, Big Ag and its allied global industries produces entirely new profiles of the American rural town. As urban and regional planning Professor Faranak Miraftab describes in her study of Beardstown, Illinois (pop. 5,925), some towns are not majority Black or white populations at all but reflect a rich, globalizing new placemaking dynamic made by migrant and transnational labor, brought on in the context of large-scale agricultural production or manufacturing (Miraftab, 2016). Caldwell, Idaho in this book celebrates an increasing Hispanic population within a similar context of migrant labor. Bentonville, Arkansas and neighboring Rogers and Springdale, Arkansas, also in this collection, are perhaps the most global towns of all. They demonstrate how service workers and executives live and work across a constellation of towns to fuel global corporate industry giants that, in turn, decide how to invest in rebuilding their landscapes, institutions and public realm.

The road to Bentonville, Arkansas is paved by Walmart, Inc.

Towns are also national, as rebuilding efforts have captured a niche market of U.S. TV and celebrity media with growing popularity. The designer-construction crafters Erin and Ben Napier market incremental residential redevelopment to a broad public in the HGTV show *Home Town*, that showcases neighborhood change in Laurel, Mississippi (pop. 17,146) and its spin-off, *Home Town Takeover* that aims to transform the entire town of Wetumpka, Alabama (pop. 8,278) (Haynes, 2020). These series follow Joanna and Chip Gaines' show *Fixer Upper*, which also began with incremental house-by-house improvements but expanded into the redevelopment of an (abandoned) silo complex and other buildings in downtown Waco, Texas (pop. 138,289). The Gaineses ultimately launched a real estate, media and retail empire called Magnolia that is visible everywhere in the city—to the point that some residents cite burn-out and ask, "When is this all too much?" (Linauer, 2022). Magnolia unabashedly leads urban redevelopment and public spending (Smith, 2017) in Waco as the city image itself bleeds into the Magnolia brand (Magnolia, 2024). Shows like this bring national focus to cities and town revitalization, as they also exemplify the extreme of neoliberalism's tendency to marketize everything, including the public realm and city-making itself, to a fault.[4] The case of Greensburg in this book, and its inspiration for a new Discovery Channel series, presents a slightly different spin on town TV, yet still highlights how the presence of film crews and the national spotlight can steer rebuilding in very real, if unconventional, ways.

Enhanced (Non-Motorized) Mobility Is More Possible in Towns

With climate change worsening from fossil fuel emissions, progressive towns stand out for their embrace of alternative healthy forms of non-motorized mobility: walking and biking. These may seem like old-school modalities to get around, but they present viable alternatives when public transit systems otherwise lack population densities to be feasible. As the case of Atlanta's Beltline

Bike parking is popular at Indian Creek Plaza in Caldwell, Idaho.
Credit: © Eleven One Photography.

and other cities' greenway and trail efforts show, retrofitting new bike and pedestrian networks into existing or new alignments can be exceptionally complicated (Gamble and Heyda, 2016). Yet small towns have been able to literally fit them in. The slower streets and reduced congestion in many smaller municipalities make them conducive for bike-based mobility. Increasingly, trails, greenways and other kinds of connections are extending into the landscapes around towns, and connecting towns with each other. Not only are town streets easier to share, but also many, like Jamestown, New York, Piqua, Ohio or Bentonville, Arkansas have turned bike culture into a burgeoning economic market.

"Programming" Has Many Dimensions

Urban design and civic projects take on much more varied meanings depending on a town's economic profile. In most major urban design and public realm projects today, like in the tourist-oriented town of North Adams, Massachusetts, programming refers to the strategies for making the downtown and its primary art institution an entertainment and leisure space. Public space programmers plan events to bring people together and stimulate economic activity. But in places like Opa-locka, Florida in this book, "programming" refers not just to leisure but to the provision of core local services. Both dimensions of the term describe how the public sector in U.S. towns and cities has experienced diminished capacity. In both cases, for leisure-based capital growth, or for basic needs, programming in the United States is led by quasi-public–private non-profit organizations. How well they meet the needs of all depends to a large degree on leadership interests and the scale of their reach. The public–private American town that relies on external programming may never be entirely democratic. Its relative success, nonetheless, is reliant on strategic relationships, formal and informal.

Town Spatial Structures

Towns are often described with simple modifiers, even as the people, places or economies on which they once relied have evolved over time. We sought to categorize the towns in this book according to typologies that avoid the limiting descriptors commonly used which are tied to landscape features, infrastructures or economies: hill town, rail town, canal town, resort town, college town, company town or mill town. A town's growth trajectory is often commonly used as a means of describing a place, like a boom town, one-horse town or ghost town. There are also modifiers of racism and control: a sundown town is where people of color were forced to leave after daylight, a description of exclusion.

Since no place cleanly fits any singular label, we developed a comparative framework *based instead on each town's spatial typology* in relation to other towns or cities around it, or to its larger region. Our typologies capture the important reality of any given town's situational political and economic context as it is inflected by its geographical stance. While not all towns within a shared, locational typology experience the same access to opportunities, they do reflect recurring patterns. The four spatial structures are described below and organize this book.

Constellations

When a few smaller towns are located near each other (and are not immediately outside of a major city), they inter-operate as a related cluster, or constellation. Although they may have arisen in isolation, towns in a constellation eventually grow closer within a geography and in ways that now influence, resist or draw from each other. These towns benefit from shared priorities and visitors as well as from common corridors and landscape features. They can also form alliances in ways that expand their collective footprint and access to resources. As the cases in this book show, a constellation of towns can be efficient and strategic. Together with adjoining municipalities, a constellation of towns becomes interdependent at a regional scale.

Satellites

When a few municipalities are located outside of, but near to, a larger town or mid-sized city, they have a satellite-type relationship to it, much like a bedroom community. Leveraging the close proximity of an urban center while maintaining their own identities, these smaller municipalities are close enough to access the benefits of a larger city and financial center while minimizing negative impacts locally.

Metroposts

Towns situated within a very large metro city-region or county are what we call metroposts. These towns are subject to increased competition for federal, state and regional resources and projects, but also may have opportunities for more diverse investors, residents and programs. For some small towns, being part of a large metro brings access to corporate investments that wouldn't otherwise happen if the town were more isolated. At the same time, if being smaller means regulating development with autonomy, it can also mean that autonomy comes with less voice in relation to powerful regional politics. At extremes, some wealthy metroposts wield exclusion to become enclaves, as poorer or demographically transitioning metropost towns risk being exploited by dominant corporate and state interests.[5] The cases in this book show how metroposts assert creative local strategies for serving their residents directly. The towns draw on local knowledge to resist, embrace or shape urban agendas in the midst of broader regional forces.

Outposts

Towns that aren't near another town, nor embedded in a larger metropolitan area are outposts. Larger than a village but smaller than a city, these towns' footprints grow independently of any particular relationship to another town. In many cases, their economies developed in harmony with the land and infrastructures around them. Many outpost towns began and matured in sync with farming, tourism or outdoor access. However, as those economies and related jobs have shifted, outposts today look for the next markets to sustain them. A theme of sustainability and conservation of resources connects the outpost cases.

Here, then, are the towns in the book, briefly described by section and typology.

Section 1: Constellations

North Adams, MA (pop. 12,924)[6]

Contemporary arts re-inhabits the shells of an industrial past, in the town and broader hills that increasingly draw tourists and outdoor enthusiasts to North Adams, Massachusetts. Credit: Douglas Mason.

North Adams, Massachusetts lies in the Berkshire Mountains in close relation to Williamstown, Massachusetts, and a few other historic hamlets with arts and leisure at their core. It is also a historic textile mill town, and the scale and materiality of its abandoned industrial buildings have made beautiful shells for new uses and forms of loft living. At the center of the town, its largest textile mill complex was repurposed into MASS MoCA, the Massachusetts Museum of Contemporary Art. This immense project helps to thread together the constellation of arts-related towns in the region. Still, until only recently, the popular museum and performance venue has operated more as an elite enclave walled off from its declining post-industrial host. A new urban design strategy re-centers both former mill/arts campus and downtown core with a new public green and infrastructural investments to overcome the disconnect.

Jamestown, NY (pop. 28,579)

Like North Adams, economic prosperity in Jamestown, New York was closely tied to early transportation; the town's fortunes rose and fell with the railroads. Also similar to Bentonville, Arkansas and others in the book, redevelopment today hinges on the generosity of its private foundations. Jamestown also found *humor in rebuilding*: It attracted the National Comedy Center that repurposed the historic train depot, negotiating a grade change in the hilly terrain, reclaiming its once industrial riverfront and putting Jamestown on the map for entertainment enthusiasts. Recreational trails ring the waterfront to concentrate and reinforce the rebuilding efforts downtown. Meanwhile, as Jamestown draws new visitors, it enters into an enhanced programming relationship with the nearby Chautauqua Institute on the other side of the shared lake. Together this constellation of two towns aims to help stabilize the Snow Belt, a water-rich region surrounding Chautauqua Lake and Lake Erie.

Bentonville, AR (pop. 54,675)

Bentonville, Arkansas lies in Northwest Arkansas, one of the fastest growing areas in the United States. This growth is partly due to Bentonville's position within a constellation of similar towns, each with a global industry or major university to anchor it: These include Springdale, Rogers, Fayetteville and a few others. The growth, however, and wealth that comes with the anchors unevenly fills the corner pocket of the state, which otherwise still experiences one of the highest poverty rates in the nation. Transformation in these towns has been largely achieved at the hands of Walmart, a multinational corporation headquartered in Bentonville, as well as by the founding family's nonprofit foundation, the Walton Family Foundation (WFF). If Walmart big box stores are infamous for killing Main Street and the small town, then in Bentonville, Walmart now remakes the town—even if on its own terms. Bentonville and the surrounding towns are not linked by public transit (there essentially is none), but by a major greenway that ties into extensive mountain biking recreational trails and other connective corridors. The constellation is an extreme example of the American system of city-making that benefits, even as it relies on, its largest corporations and wealthiest actors.

Section 2: Satellites

Caldwell, ID (pop. 60,482)

Caldwell, Idaho is located west of the state capitol Boise, in a fertile agricultural region considered a part of the "Big Ag" industrial economy and a budding wine industry. A large migrant Hispanic population grows or shrinks with the harvest to fuel these industries. Caldwell's affordability also draws on its close proximity to Boise, where many commute daily in exchange for small town living. Despite the influxes, Latino and white residents still largely reside on different sides of town. Meanwhile, over the course of nearly two decades, Caldwell has systematically uncovered its forgotten buried creek, a project that also allowed the town to establish a new armature of spaces to bring together these diverse publics. The rebuilding strategy restores the waterway to a naturalized condition with trails and arts-inspired bridges, anchored by a new offset public square. These urban elements stake new space of co-belonging and co-ownership for the mix of people and businesses in the Intermountain West.

Piqua, OH (pop. 20,368)

Piqua, Ohio was built at a crossroads of trails, rails and roads that facilitated trade and connections to Dayton twenty-eight miles (45 km) south. Its oldest infrastructural system, however, is what now anchors the town's contemporary transformation. A 19th century limestone canal sliced through the downtown and, like Caldwell's Indian Creek, has been excavated with its path re-established. Uncovered after a century, the corridor anchors a new waterfront park designed to work with future flood events. The park reorients a public realm network and repositions the former industrial buildings along its path. It serves as a trailhead to 330 miles of continuous paved trails along the Great Miami River whose recreational use has been a major factor in its resurgence connecting the Piqua satellite to Dayton and the larger Ohio and Indiana region.

The old lock of the future Lock Nine Park under construction in Piqua, Ohio just north of Dayton. Credit: Greg Merrick/Choice One Engineering.

Section 3: Metroposts

San Ysidro, CA (pop. 28,008)

As a *metropost*, San Ysidro lies in the shadow of San Diego, California.[7] It is technically an outlier "town" since in 1957 the unincorporated community was annexed into the city of San Diego, its massive urban neighbor to the north. It appears in this book because it represents a long-standing community with a strong identity relative to the big metro, but also under yet a larger umbrella: the global border condition between the United States and Mexico. Radically different scales of design, and socioeconomic priority, inflect this small suburb. The speed and scale of moving a tremendous amount of goods and people through the border at San Ysidro signals its role as an important center—or pinch point—of global economic flows. On the ground a few blocks from the border, a local non-profit has inserted contemporary community-based services to serve the population. Casa Familiar is a reminder that San Ysidro is host to local space and local people, their stories and needs. The case tracks scales of urbanism across economies, borders and constituents; global to particular.

Opa-locka, FL (pop. 16,357)

Opa-locka, Florida is a living gallery of a 1920s concept for a Moorish-Revival-architecture themed town that offered an escape out from under the umbrella of downtown Miami just ten miles away, even as both municipalities still exist in the larger shadow of Miami-Dade County. Today, at least twenty of the Moorish buildings are listed on the National Register of Historic Places and the town's street names still reflect ideals of its North

The Moorish-Revival City Hall in Opa-locka, Florida.

African guise. Over the course of time, the town's identity became fused with a majority African American constituency that today celebrates African and African American art and cultures. Opa-locka has become an emerging center of Black contemporary arts. But years of broader economic and policy exploitation maintain it as one of the poorest municipalities in Miami-Dade County. The metro region's wealthiest actors and corporations access Opa-locka's private airport and affiliated tax incentives while the town still awaits crucial infrastructure and other services. As affordable housing becomes scarce everywhere, and as sea-level rise threatens Miami, this inland town now faces double gentrification that is economic and climate-driven. Here is where the Opa-locka Community Development Corporation (OLCDC, renamed Ten North Group in 2022) builds on its forty-plus years of local support with the long view in sight. In the context of a weakened public sector, uneven global wealth and with a rich connection to contemporary art, urbanism by this CDC, like in San Ysidro above, has been incremental but vital, varied and creative, keeping people first.

Section 4: Outposts

Traverse City, MI (pop. 15,501)

The Traverse City case highlights challenges associated with the preservation and adaptive reuse of large historic complexes and landscapes, underscoring the tensions between growth and land conservation, locals and visitors, tourism and livability. In a remote but fertile region surprisingly connected with seasonal direct flights from many major U.S. cities, this outpost town demonstrates an enlightened regional planning perspective based on sustainable principles that redirect rebuilding away from precious resources and towards the downtown. In that context, a monumental, historic hospital campus was adaptively reused for contemporary needs, including housing across the affordability spectrum. The historic structures create the building blocks of revitalization apart of the actual town core. The large redevelopment models climate-appropriate adaptive reuse that serves a growing population without treading on the land.

Greensburg, KS (pop. 732)

Greensburg, by far the smallest town in the book, is an outpost in southwestern Kansas on the Great Plains. The hamlet of 1,400 people was losing population when it gained national attention after being obliterated by a tornado in 2007. The Category 5 storm destroyed 95 percent of the town, calling into question the viability of rebuilding at all. However, the community of farmers, long connected to the land around them, rallied to rebuild and to do so as green as possible. In a political era of stark divisions that pit rural against urban, conservative against progressive, Greensburg melded them all. The story of its rebuilding is one of progressive sustainable design that strengthens the town's traditionalist connections to its natural resources.

Small town Main Street in Greensburg, Kansas is also a water conservation system with native plants and rain gardens.
Credit: © Assassi/BNIM.

The peripheral outpost briefly became a national confluence of diverse groups from around the region and country who came to rebuild, including the actor Leonardo DiCaprio and the Discovery Channel. **Greensburg models a political path for rebuilding the American town that sheds red versus blue ideologies, for green**.

Methodology *Defining Rebuilding*

The cases in this book underscore the range of challenges and opportunities inherent in repositioning towns relative to global, national and economic markets. To highlight specific urban design strategies and features, each case includes a descriptive map and perspective 3d diagram. Our inclusion of these graphics aims to situate the complexity of each project in the broader physical and political context of the place. The drawings follow a consistent format to enable comparisons across scales. Many cases also include an information graphic, to highlight comparative data or to accentuate an aspect of the project's evolution or impact. Additionally, each case includes a *Perspectives* section that highlights the voices and stories of those involved in the projects described. These interviews greatly enriched our understanding of the complexities and strategies (formal and informal) for rebuilding. The book and cases intentionally traverse a variety of project themes and convey a range of actors, scales and locations across the United States to test the limits of any given paradigm, and to illustrate as many rebuilding lessons as possible.

As with *Rebuilding the American City*, this book avoids rating the projects it highlights in terms of revitalization "success," as it became clear during the course of research that success was a relative and subjective value. More often than not, success is measured financially, and does not always include the most critical environmental and social health of a community or its immediate needs. By drawing on perspectives from constituents and civic leaders, designers, public officials and private developers, the book highlights the tensions inherent in achieving a balanced regeneration within a given context.

Instead of narrowly defining *rebuilding* as a set of singular assumed redevelopment goals that blindly apply to each town, we sought a more layered approach. Each project's immediate issues—and typological stance—informed the analysis and our framing of design and social impacts of the projects. The book intentionally seeks to unravel the urban design interventions and associated planning strategies in such a way as to re-position them as nuanced, idiosyncratic combinations of responses to the difficult questions of rebuilding: *What does contemporary development in the small American town mean, and for whom? How is it achieved, and what does that look and feel like?*

Notes

1. There is no universally accepted definition of a town. We recognize that for some people a town of 50,000 people would be considered a small city and that some towns of less than 20,000 people define themselves as a city. The use of the term town varies by state, its type of governmental structure and whether it is incorporated or unincorporated.
2. The homelessness among individuals rose by nearly 11 percent; among veterans by 7.4 percent and among families with children by 15.5 percent. The statistics are worse for people of color. Those who identify as Black make up just 13 percent of the U.S. population but comprised 37 percent of all people experiencing homelessness. See https://www.boston.com/real-estate/real-estate-news/2023/12/15/us-homelessness-up-12-percent/?s_campaign=Email:Address.
3. All population figures are from 2020 (U.S. Census Bureau).
4. To understand the extent of the marketization of all things around us, and the deeper social and cultural ramifications of neoliberal political economic ideology, see, for example, Brown (2017).
5. These extremes become much more visible in highly fragmented metro areas like St. Louis, Missouri where over 88 separate towns exist under a regional umbrella and thus compete for—and exploit in racialized ways—access to corporate investment and resources (see Heyda, 2024).
6. All population figures in this section are from the year 2020; U.S. Census Bureau unless otherwise noted.
7. San Ysidro population is from Wikipedia for 2010. See https://en.wikipedia.org/wiki/San_Ysidro,_San_Diego.

References

Brown, W. (2017). *Undoing the Demos: Neoliberalism's Stealth Revolution*. Brooklyn: Zone Books.

Campoli, J. and MacLean, A. (2007). *Visualizing Density*. Cambridge: Lincoln Institute of Land Policy.

Flavelle, C. (2021, September 2). Climate Change Is Bankrupting America's Small Towns. *New York Times*. Retrived from www.nytimes.com/2021/09/02/climate/climate-towns-bankruptcy.html

Fox, J. (2020, August 27). The Rise of Work-From-Home Towns. *Bloomberg*. Retrieved from www.bloomberg.com/view/articles/2020-08-27/scenic-towns-enjoy-boom-as-work-from-home-becomes-pandemic-norm

Gamble, D. (2023). *Idea City: How to Make Boston More Livable, Equitable and Resilient*. Amherst: University of Massachusetts Press.

Gamble, D. and Heyda, P. (2016). *Rebuilding the American City*. New York: Routledge.

Haynes, B.L. (2020, July 2). Ben and Erin Napier Announce the Town Selected for New HGTV Series 'Home Town Takeover'. Retrieved from www.hgtv.com/shows/home-town/ben-and-erin-napier-announce-town-selected-for-new-hgtv-series-home-town-takeover

Heyda, P. (2024). *Radical Atlas of Ferguson, USA*. Lakewood: Belt Publishers.

Hickle, J. (2020). *Less is More: How Degrowth Will Save the World*. London: Penguin Books.

Kelly, J.P. (2023, November 22). Small-Town America's Never-Ending Struggle to Maintain Its Values Hasn't Always Been Good for US Democracy. *The Conversation*. Retrieved from https://theconversation.com/small-town-americas-never-ending-struggle-to-maintain-its-values-hasnt-always-been-good-for-us-democracy-213957?utm_medium=email&utm_campaign=Daily%20Newsletter%20%20November%2022%202023%20-%202802228359&utm_content=Daily%20New

Linauer, K. (2022, July 14). This Is Why So Many Wacoans Struggle to Celebrate Magnolia (Fixer Upper) & Their Impact On Waco. *Addicted to Decorating*. Retrieved from www.addicted2decorating.com/this-is-why-so-many-wacoans-struggle-to-celebrate-magnolia-their-impact-on-waco.html

Lipsitz, G. (2007). The Racialization of Space and the Spatialization of Race: Theorizing the Hidden Architecture of Landscape. *Landscape Journal*, 26(1): 10–23.

Loos, T. (2017, August 18). Columbus, Ind., Renews Its Big Design Legacy. *New York Times*. Retrieved from www.nytimes.com/2017/08/18/arts/design/columbus-ind-renews-its-big-design-legacy.html?mcubz=3

Magnolia (2024). *Come See Us in Waco*. Retrieved from https://magnolia.com/visit/

Means, M. (2020). *Main Street's Comeback*. Silver Spring, MD: Hammondwood Press.

Miraftab, F. (2016). *Global Heartland: Displaced Labor, Transnational Lives & Local Placemaking*. Bloomington: Indiana University Press.

Orvell, M. (2012). *The Death and Life of Main Street*. Chapel Hill: The University of North Carolina Press.

Smith, J.B. (2017, June 9). Magnolia's 6th Street Sidewalk Project to be Done by Next Month. *Waco Tribune-Herald*. Retrieved from https://wacotrib.com/business/magnolias-6th-street-sidewalk-project-to-be-done-by-next-month/article_f9f18628-6d84-51ca-9907-e31e7b4a5c1d.html

Section 1 **Constellations**

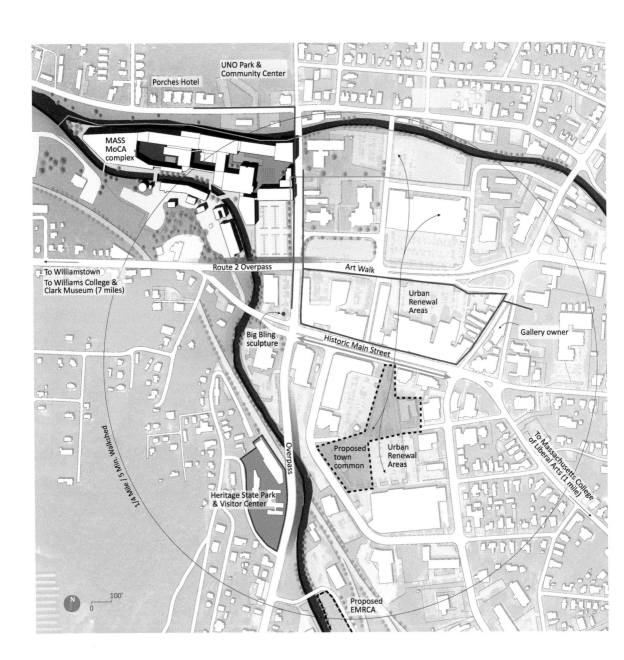

1.1
North Adams, MA

North Adams is a town of 12,880 residents[1] located three hours by car from Boston or Brooklyn. The smallest "city" in Massachusetts is nestled in the western highlands of the Berkshire Mountains.[2] It sits in what we call a constellation with Williamstown, Massachusetts seven miles to the west and a few others a short drive away. All of the towns in this constellation are anchored by major arts-related institutions.

Economic Shifts

North Adams, like its neighbors, was once bigger and busier. During the industrial period in the earlier 1800s and into the turn of the last century in the United States, the center of the town's economic and physical core was marked by a massive textile mill called the Arnold Print Works. By the 1830s, some fifty or more mills were located all along the Hoosic River in this bucolic part of New England. The immense mill was centrally sited—not necessarily to participate in the life of the downtown core—but to harness the confluence of the north and south branches of the Hoosic River and the network of rail lines located there.

Arnold Print Works supplied cloth for the Union Army during the Civil War, and at one point operated within 700,000 square feet of space, with 3,200 employees, spanning 28 different buildings on 16 acres (Sprague, 2016). When the mill industry declined and the company moved out in 1942, another major industry, Sprague Electric, quickly moved in to manufacture high-tech electronics. Through these two successive industrial job centers, the town's population and economy expanded. Like in other company towns of that era, large multi-unit complexes supported worker housing all around the mill, many of which are still visible in the hilly slopes defining the town. But as manufacturing economies moved south and overseas and as technology changed, by the 1970s the electric company, too, moved on.

For much of the latter half of the 20th century the town hemorrhaged population. In the early 1980s and 1990s, however, broad political economic shifts toward privatization in U.S. urban policy and redevelopment laid a path for new directions for the vacant mill and the struggling small town. One characteristic of this political economic shift was the increasingly accepted idea of state funding to support the arts—as an instigator for economic development. The empty and deteriorating mill complex became a focus of renewed possibility.

Route 2, a narrow street leading into North Adams, Massachusetts between old brick factory buildings.

A group of entrepreneurial institutional leaders from across the constellation of towns initiated a process to convert the former factory buildings into what is now one of the largest contemporary art museums in the world, called the Massachusetts Museum of Contemporary Art (MASS MoCA). The vast institution illustrates the shift that many American cities went through as they moved from a reliance on the manufacturing sector to culture- and tourist-oriented markets. **Rebuilding in North Adams today presents a focus on the arts as a driver of urban transformation—albeit one that requires having and leveraging local institutions**. While the constellation works as an effective attraction bringing tourists to the region and growing the leisure industries, it is hard to determine how well the town itself has benefitted. North Adams still struggles with vacancy and a lack of connectivity within its downtown. The case presents questions about how and for whom rebuilding occurs in the American small town.

Dimensions of Rebuilding

Like many American downtowns, North Adams also fell victim to strategies of urban renewal during the 1960s. Several blocks were cleared and massive new road over-passes built that dwarfed even the old mill itself. Clearings decimated the downtown's domestic scale and created wide riffs through it. Route 2 and State Street became concrete highway-like bridges over the railroad lines and river on the west and south sides of town. These cut the downtown core off from the mill on the west, and from other neighborhoods and destinations to the south.

The central portion of the downtown's well-defined streets with ornate, historic buildings were demolished during urban renewal and the large

The MASS MoCA complex at 87 Marshall Street seen from the confluence of the north and south branches of the Hoosic River.
Credit: © Michael Moran/OTTO.

"development" sites that resulted never attracted the imagined projects. Some were replaced instead with single-story strip malls set back from the streets by expansive surface parking lots. The town did not thrive in the years following. Instead, on the periphery, the stock of densely packed homes circling the downtown deteriorated in sync with the declining economic base. Predictably, like in many American towns and cities, transportation policies had privileged narrow notions of "efficiency" to transform the historic Mohawk Trail adjacent to MASS MoCA from a "Scenic Byway" into the fast-moving, elevated Route 2 highway, forming a nearly impenetrable edge between the former factory complex and the commercial center that it once served.[3] While the distance between the two are much closer than they seem, these car-induced barriers continue to stymie the potential of MASS MoCA to more effectively "connect" North Adams.

Critical Mass

The flexibility of the former mill spaces enabled MASS MoCA to exhibit art at a scale few peer institutions could rival. But the promised catalytic effects of the institution on the surrounding downtown remained elusive. By the early 2010s, while the internal attendance increased and the museum became a global destination, physical impacts on the town lagged and local leaders commissioned an urban design study to draw on the institution's allure.

By 2012, the North Adams Partnership was formed. This was a group of civic leaders, philanthropists and public officials who came together with MASS MoCA to consult with the New York-based economic development firm HR&A Advisors and design practice SHoP Architects to complete an economic and urban design study for North Adams. One of the primary objectives of the study was to find ways to overcome the fragmentation caused by urban renewal and to move art and performances beyond the walls of the museum and into the fabric of downtown. The master plan proposed new buildings, open spaces and wayfinding to knit the museum—its visitors— more into the fabric of the downtown. To date, multiple murals, light displays and large three-dimensional sculptures were built, most within a five-minute walk from MASS MoCA's front door. These connectivity beacons conceptually link the downtown to the institution and facilitate visual continuity and commercial activity between them. The installations operate as art-based forms of wayfinding through a variety of means and scales, linking destinations downtown along an Art Walk.

Effective urban design wayfinding is for both orientation and educational benefit, but in our contemporary public/private era that conflates cities with markets, wayfinding gets used for branding and imaging and less for public space cohesion. Since its opening in 1999, more visitors *are* venturing outside the MASS MoCA gates, but there are two other key components of the plan that would improve local public space connections and have yet to be completed. One is a proposal to remove the elevated Route 2 infrastructural barrier to bring the road back down to grade as a graceful boulevard condition. The other component is the creation of a major new green space in the heart of the downtown where urban renewal eradicated so many late 19th and early 20th century buildings. Conceptualized as a formal Town Common, the open space would remove acres of impervious pavement from the surface parking lots and stimulate new development on its edges. The master plan seeks to mend the downtown gaps. It also suggests strengthening relationships across the southern overpass-chasm, to better connect to an older tourist destination called Heritage Park on the other side of the river and train tracks.

The Western Gateway Heritage State Park Museum and Visitor Center is an earlier vision of an adaptive reuse-tourist district in North Adams on an old rail yard with buildings on the National Register of Historic Places. It was created in the 1980s to tell the story of the construction of the Hoosic Tunnel

Opened in 1986, the Heritage State Park Museum and Visitor Center (left) is close to the center of North Adams, but spatially severed by the State Street overpass (right).

(1851–1875), a 4.75 mile (7.6 km) engineering feat that brought rail to the region. Although the buildings and landscape are well maintained, today the site is mostly vacant. But following the success of MASS MoCA, yet another vision has emerged—by Thomas Krens, one of the initial MASS MoCA champions—for an immense model railroad museum just east of Heritage Park along the rail line.

The proposed Extreme Model Railroad and Contemporary Architecture Museum (EMRCA) is envisioned as a sleek new building that would house over a thousand architectural models and run more than 100 model railroad trains within its main hall. Part exhibition hall and part theme park, this new attraction is cast as an anchor that will shore up the adjacent Heritage Park. Development plans are predictably tourist-oriented. They include a luxury spa and hotel and other cultural attractions.

While the ambitions of the overall master plan (the removal of the overpass and construction of the Town Common) would clearly benefit these additional ambitious developments with easier connections, they signal a more important message to residents: That the town could and should be more than a sum of its tourist-parts. A central common demonstrates investment for residents, with better ways to get there. It also advances an environmental priority, since greening the downtown will help diminish heat island effects and the run-off impacts of so much impervious pavement.

In the interim, there is some evidence of incremental change that has spun off of MASS MoCA's momentum. Across from the museum itself, a small community center was fashioned from a dilapidated tavern, along with improvements to an underutilized corner lot turned into "Uno Park." But the majority of the museum's spin-off redevelopment impacts remain catered to high end visitors to the museum and region. Vacant Victorian houses across the street from MASS MoCA on River Street were saved from demolition and redeveloped into an upscale hotel called the Porches Inn. Other boutique hotels downtown and in the direction of Williamstown have followed. Occupied houses right next door, however, show signs of wear—local life and care—but await reinvestment.

In nearly every town and city, the allure of a "silver bullet" to save downtown is compelling, a simple and magical solution to a complicated problem. Throughout its history, North Adams has both benefited and suffered from

The popular Porches Inn (left) and local residents' porches (right) along the same street directly across from MASS MoCA.

1.1 North Adams, MA

the consequences of relying too heavily on a single industry that inevitably shifts over time as manufacturing and market forces ensue. And yet, the downtown plan would be meaningless without MASS MoCA. If the project never emerged, the complex would have certainly been demolished and the town could have rapidly declined. The importance of the institution cannot be overstated. Still, as North Adams looks towards its future, it may be wise to guard against relying on one market sector. The COVID-19 pandemic illustrated how vulnerable places can be when people stop visiting. Tourism builds transient populations over local ones. It might be a welcome byproduct of diverse economic drivers but urbanism shouldn't be about attractions per se but about inscribing a stronger sense of place. The key is to ensure that redevelopment brings meaning and wellbeing for residents of North Adams, and not a mere boutique version of a town for visitors who can afford it.

Nadine Berger, architect with the firm SHoP who worked with HR&A advisors on the downtown plan talks about local linkages that are just as important as the connections to tourist attractions: "I think small-scale interventions are more important in towns like these. We call it the 'broken links everywhere' problem. It's creating connections and tying it all together that seems to be more important for a small town than big moves. MASS MoCA was one big move, that's enough" (Berger, 2022). The downtown plan, with its connective public green common in the heart of multiple fragmented blocks, is not yet realized, but it could be the most meaningful of all the efforts so far in North Adams to leverage the investments in MASS MoCA that put North Adams on the map. The strategy creates spatial cohesion downtown, and provides a more public-oriented plan than the new ideas for more attractions on the south side.

Berkshire Constellation

If internal spaces of North Adams remain fragmented, the external relationships with towns around it are thriving. MASS MoCA exists within an ecosystem of a number of other prominent arts and cultural attractions nearby, including Jacobs Pillow in Beckett (an outdoor performing arts and education center), Tanglewood in Lenox (a music venue and summer home of the Boston Symphony Orchestra), and the Clark Art Institute in nearby Williamstown, whose sylvan campus is comprised of numerous buildings from celebrated architects and landscape architects.[4] Far from being an island, then, the art and music destination was always envisioned within an archipelago of cultural, artistic and educational venues quite close to one another, including numerous institutions of higher learning. And while college students are not tourists, they represent a transient population. Still, colleges enliven towns with student life, faculty and staff, and with buildings and programs open to the public. In addition to nearby Williams College, the region benefits from 30,000 students from the other "Five Colleges," plus the Massachusetts College of Liberal Arts (MCLA) in North Adams. These institutions bring a constant influx of youth and energy into Hampshire County.[5]

Williams College is the closest, a mere seven miles down the Scenic Byway to the west. A small private liberal arts school with 2,000 students and over

300 faculty and staff, the college brings attention and resources to nearby Williamstown. It is no mistake that it was three alumni and employees of Williams who initially spearheaded the MASS MoCA project, and that the largest benefactor/philanthropist also hails from the school. Many American cities experience redevelopment by way of the capacities of their major institutions—universities, hospitals, museums and other intellectual centers. For smaller towns, North Adams shows how this is still the case and an established model for expanding one's institutional footprint while also shoring up neighboring physical environments. The effort incentivizes the development of additional amenities, like hotels and attractions for parents and visitors to campus.

The common urban design paradigm that leverages institutional anchors for redevelopment now operates across multiple allied jurisdictions in this constellational model. The move to house art in North Adams just down the road from Williamstown shows how institutions in small town settings are conceptualizing and expanding beyond their single campus "footprint" into multiple spatially dispersed footprints that traverse town lines. These expansions reinforce the institutions' individual identities while collectively claiming a larger coordinated regional presence.

MASS MoCA began as Williamstown was initially looking to move some of the larger art works of the school's collection into an inexpensive, interior environment that could accommodate them. From there, the plan flourished into a much larger vision that could take advantage of the empty industrial

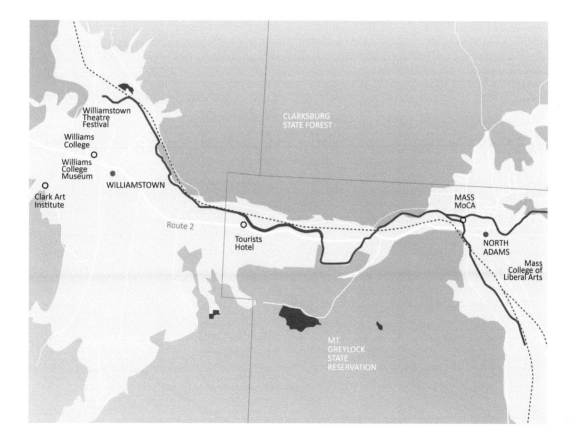

1.1 North Adams, MA

shell only a few miles away in the adjoining municipality. The former mill complex was close enough to visit by car—and even via bike and hiking trails—and provided a way to initiate improvements on what was otherwise a declining urban area close to the college.

Similar to the Bentonville case in this volume, the identity of North Adams as a town can be attributed, at least in part, to the tourist amenities that surround it in the region. This is not necessarily a new concept. Paradoxically, the remoteness of the Berkshire region has long made it a place for tourism. Opening in 1914 as one of the country's first auto-touring roads, the Mohawk Trail (now Route 2) is an official Scenic Byway that runs through Massachusetts from New York. Many existing cultural institutions are along it, as well as a network of picturesque rivers and streams. As car ownership increased in the 20th century, motels sprang up along it which also provided access to the region's recreational amenities. As more urban hotels emerge or are being renovated around MASS MoCA, additional similar venues are being imagined in the nearby constellation. The Tourists is one such venue built in the shell of an old motor inn midway between North Adams and Williamstown. That investment benefits from visitors to the arts, the college and the woods and trails that link them. The hip hotel was co-created by developer Ben Svenson and John Stirratt, the bassist from the band Wilco, who plays every other year at MASS MoCA and is originally from the area. The partnership represents yet another feedback loop of mutually growing wealth between the arts programming at MASS MoCA and surrounding amenities—John Stirratt doesn't just stay at the hotel, he co-owns it.

Global Local Relationships

The museum is in a small town, but its formulation is big, in time, resources and private actors. The three key individuals responsible for the effort would go on to prominent roles in the international art scene. Tom Krens became the Director of the Solomon R. Guggenheim Foundation in New York, opening, among other things, the Guggenheim Museum in Bilbao. Michael Govan went on to establish Dia:Beacon in New York's Hudson Valley and Joseph Thompson became MASS MoCA's first Executive Director, a position he would hold for thirty-two years, eventually stepping down in 2022. The planning for the property was equally big. Initial master plans for the site garnered attention from world-renowned architects with a winning proposal submitted by Frank Gehry, Robert Venturi, Skidmore, Owings & Merrill and Simeon Bruner of Boston-based Bruner/Cott, who would go on to become the primary architects for the project's implementation.

The rehabilitation of MASS MoCA was enabled by the three-decades-long public–private partnership involving state, regional and local government, many federal agencies and private-sector individuals who, at key moments in the institution's evolution, challenged it to do more to raise funds for the operations and maintenance of the enterprise. An initial influx of capital in the form of a $35 million expenditure from the Massachusetts state legislature eventually emerged, but it took years for the funds to be allocated as administrations changed and the economic climate oscillated. Eventually, the rehabilitation and repositioning of 87 Marshall Street would span 30 years, three mayors, five senators and seven governors.

Developing Arts

MASS MoCA stands out as one of the earlier examples of the paradigm of as economic development that can be witnessed today around the country. Bentonville and Opa-locka in this volume provide additional and different examples of how art is leveraged to stimulate rebuilding. West Main Street in Louisville, Kentucky (highlighted in *Rebuilding the American City*) was another effort to first stabilize, then rebuild an historic downtown corridor on the premise of art and the culture of fabrication.

The MASS MoCA art concept has grown from one of large-scale static exhibitions to a multi-disciplinary center for visual, performing and media arts. The complex has also cultivated a unique reputation for "making," fostering the production of new works that are not possible elsewhere. Beyond art, approximately one third of the space at the complex is dedicated to revenue-generating commercial tenants. This provides an income stream for the museum's bottom line. The dynamic of a constantly changing performing arts calendar, together with the other venues that include a bar and restaurant, brings people back to MASS MoCA at all times of the year. If the paradigm is about towns bringing arts institutions in to produce trickle-down spin-off development, then MASS MoCA shows another cautionary dimension of that idea, where the arts institution itself *becomes* the development. Do the arts in this case spur a remaking of the public town or does the arts center become a new privatized version of public life instead?

MASS MoCA investor and philanthropist Jack Wadsworth said:

> For twenty years—the proof is in the pudding— MASS MoCA in North Adams did not work. I mean, it was a failure! It took unemployment from 25% to 10%, but it stuck at 10, and population continued to decline. The schools got worse. Real estate prices went down. The city was financially strapped. You know, people came to MASS MoCA in droves, but they got back in their car and they left.
>
> (Wadsworth, 2018)

He was speaking in relation to the broader promised impact of the museum on town conditions. While MASS MoCA continues to grow and spin-off *luxury* amenities for visitors, it remains to be seen how that wealth might become more legible in the city for residents. While non-profit institutions don't contribute taxes to a city, hotels, new housing and other developments do. New hotels continue to be built or renovated in North Adams.

Subtraction as Strategy

At the scale of the complex itself, the buildings at MASS MoCA were given an attention to detail with a respect for the craft of architectural

Building uses change but the forms endure. Image of one of the large open spaces in the mill complex initially designed for textiles and then electronic components before its renovation into a gallery for rotating contemporary art. Credit: Bruner/Cott.

intervention that the town of North Adams itself was seldom since afforded. The large spaces of the former Arnold Print Works and Sprague Electric complex lent themselves well for conversion to art galleries in terms of their high volume and expansive floor plans. However, the benefits of flexibility for production were offset by the challenges posed by the relentless pattern of fenestration. The 2,500 windows that brought light into the interior of the buildings made it difficult to exhibit works that were vulnerable to direct sunlight. The architectural strategy therefore became one of *subtraction*.

Bruner/Cott, the firm that was hired to design the project following the master plan feasibility study, built out interior spaces for highly specialized rooms. Specific works by James Turrell, for example, needed to entirely block out natural light. In other cases, the firm carved out voids to bring in natural light, exposing other areas. With a very limited budget, the design intervenes in ways both visible and invisible, often blurring distinctions between new and old.[6] Rather than sanitizing the spaces and painting the surfaces anew, traces of the past are frequently evident in the floors, walls and ceiling.

Unlike in the downtown during urban renewal, most of the buildings at MASS MoCA were preserved. The structures that were in the worst condition

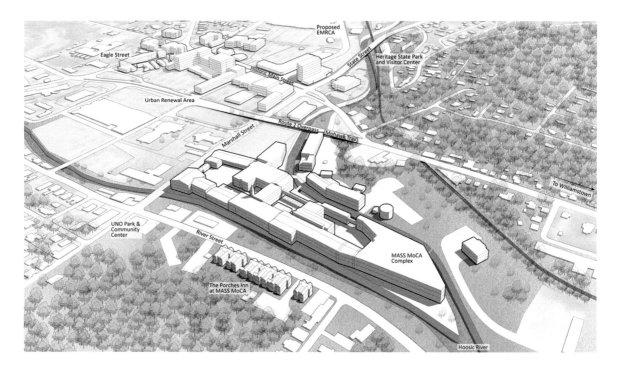

were stabilized first so they wouldn't collapse. Selective demolition brought visitors deep into the complex at the museum's entry point marked by an iconic clock tower and plaza court. Visitors follow a sequence of interconnected spaces. The complex has great urbanism; one can openly wander for hours through the campus, rambling between buildings and crisscrossing courtyards. The pathways are intuitive, and one is guided less by signage than by natural light. New and old are delicately interwoven. The project prompts the question, is this kind of care of design and spatial continuity, materiality and sequencing only possible in private investments? Or can that quality of design and standard of intervention guide the rebuilding of the town?

1.1 North Adams, MA

Perspectives

Michael Dukakis (D), Governor of Massachusetts from 1975 to 1979 and 1983 to 1991

When I started out as Governor in 1975, Massachusetts was a basket case, with the second highest unemployment rate in the country and proportionally the biggest state deficit. I'm looking at a sky-high unemployment rate generally, and real despair in these communities. So, I walked into this with this notion that, first of all, we were going to put together an economic development strategy, which the state didn't have. And secondly, it was going to focus on the 20 or 25 older industrial cities that were in terrible shape because their industries were gone. North Adams happened to be one of these—relatively small, but had been a railroad center and some textiles. They were bereft.

I got a call from John DeRosa, who said, "I want you to meet a guy by the name of Tom Krens who is the Director of the Williams College Art Museum, because he's got some ideas about what to do out here." Me and DeRosa arrive at the Williams College Art Museum and this guy proceeds to tell me this preposterous idea for the empty Sprague Mill. He spins out that it would be possible to turn the mill in North Adams into a world-renowned Museum of Contemporary Art.

Now first, I am not a big fan of contemporary art. And secondly, I'm saying to myself, "I don't know how this is going to play back at the State House."

The Mohawk Theater on Main Street in North Adams, Massachusetts in 2022.

But I listened, and came back and met with Al Raine, who was my chief of development. After a couple of weeks he comes back to me and says, "We think this will work but it will take a $35 million state bond issue." Now, this was the mid-eighties, and the state was in the Massachusetts Miracle. We had solved a lot of our fiscal problems, we had a few bucks. But a $35 million bond issue?

I first went to the Senate President Bill Belcher. Belcher is a guy that goes to Tanglewood on a regular basis; he was a member of the BSO (Boston Symphony Orchestra) Corporation. He said "Fine, yeah, we can do that." Then I went to Tom McGee who was the Speaker. He was a tough Irishman from Lynn, an ex-Marine. How is he going to react to a $35 million bond issue? And I discovered that McGee was secretly a fan of dance. I think he used to sneak out to Jacob's Pillow periodically. He said "Fine." So we start going to work and begin to put things together. I'm not sure if people were excited about this. I mean, contemporary art in North Adams, Massachusetts? But we worked at it and kept moving. I left office and Bill Weld succeeded me, and there was another recession. Weld—by the way, he's a good friend; we practiced law together before we became Governors—was quite skeptical. I said to him "Look, this isn't arts and crafts, this is economic development. Northern Berkshire is going to die as an economic entity if we don't do something." And I pointed out to him that this was the only thing that I'd seen, however unlikely, coming down the pike that seemed to be real. Weld was very concerned, but a lot of folks went to work and convinced him that over time this really was the one thing that might revive the economy in northern Berkshire County.

Kate Collignon, Managing Partner, HR+A, Advisors for Downtown North Adams Master Plan

The major economic anchor in North Adams is MASS MoCA. Ideally, you would also be proximate to a primary marketplace where you could build a supporting economy, but that's not present here because you're too far away from the major urban centers. But then you also really need strong civic leadership and that's what the North Adams Partnership was trying to create. There are all these different governmental, non-profits, civic leaders, philanthropic interests here, who are all committed to making something happen. How do they build the capacity, given the fact that they all have their own day jobs?

. . .

Almost all the places where we do work have a philanthropic base, and that's often what's driving the focus on reinvestment there. That's where a lot of the leadership ends up coming in, and where those additional resources are. Because in order to tap those public programs, you need people who are dedicated to seeking them out, and for small municipalities, particularly those that are already struggling fiscally, there's only so much capacity to put together creative ideas and put muscle around things. So, philanthropy usually does play a really big role. Jack Wadsworth [local investor] is fairly unique in his philanthropic relationship with North Adams. Usually, we're seeing a community foundation that is connected to a legacy Fortune 500 company that has historic roots in the community. You also have to get the public sector involved because once you actually start to build momentum, you want to make sure that what's being built is going to serve local residents. Public sector involvement also avoids conflict when inevitably you deal with regulatory approvals and permitting. You want those partnerships created ahead of time. And it also gives you the ability to tap into those broader public funding sources needed to round out the philanthropic contributions.

Every plan for every small community should be looking at building on the needs and resources that the existing population brings to bear. That often means pulling in resources from outside the community in addition to those who are going to be spending locally. That's where the tourism comes in, but unless that is directly feeding into local residents, you're not ultimately achieving revitalization goals. There is a lot of nuance around how much is directly serving local residents, how you are bringing in external spending, how that spending then makes its way into the tax base, and then how that ends up supporting local service distribution. All of those components usually need to be there, but not every plan ends up focusing on each one.

Jack Wadsworth, Investor/Philanthropist

A group of us created the North Adams Partnership out of which came the urban plan. In the first five years of the urban plan nothing much was done. The city had no planning department to speak of, so the mayor was almost without staff, and the North Adams Partnership had some staff but no authority.

. . .

One of my favorite snippets is that the feasibility study that was written to persuade the State of Massachusetts to part with $35 million for the original funding of MASS MoCA, was used—barely marked up—by the Basque Government in Bilbao for the Guggenheim, which took off like a rocket. So ironically the idea that museums can foster economic growth was *invented* in North Adams.

One of the things that came out of the urban plan was an initiative to attract new companies to North Adams. Tourists, an old restored motel, is a joint venture between a real estate character from Boston and the bass player from Wilco. I mean, why not? The Porches we built and own. In order to be able to call it the Porches at MASS MoCA we comp'd them 365 rooms. For the first 15 years they didn't use them and occupancy was at 30 percent. This year [2018] they were out of rooms by March 31.

. . .

This had nothing to do with confidence. It was an act of philanthropy. Pure philanthropy. We never owned a hotel before in our lives. I said to Joe [Thompson] at the opening of MASS MoCA, looking out the window at the vacant houses across the river, "That is the worst neighborhood in North Adams." But the buildings were also architecturally sort of interesting because they were Victorian wood-frame houses built in the 1850s and 60s for factory foremen. I said, "You're going to have to do something about that block," and

View of Uno Park and Community Center, previously a derelict tavern and vacant field and now a community-based space adjacent to the museum on River Street. Other residual spaces adjacent to the complex remain walkable, urban wilds.

Joe said, "Why don't you?" and I said, "I'll fund it, you do it," and we had a handshake at the window. A month later I owned the whole block. MASS MoCA had opened phase one and there were 40 workmen who were all local and had worked on that part of the museum, so the day it opened the contractors had nothing to do. They moved across the street and did the Porches. I just sent money and Joe managed the whole process, on time and on budget.

Nancy Fitzpatrick, whose family owns the Red Lion Inn in Stockbridge and was a board member of MASS MoCA, had called me on the phone and said, "I heard you bought the block across the street. How about a hotel?" And I said, "Nancy, I've got to tell you that's the dumbest idea I've ever heard in my life." Even dumber than buying the block. But six months later she came down to New York with a mission statement, a cash flow forecast, a draft of a contract, and we made a deal in a half-hour.

Step one was to raise $800,000 to pay of all the accounts payable. You weren't going to be able to keep the lights on for more than a week or so if you didn't do that. Then step 2 was to raise money to get to the beginning of their fiscal year to pay for the deficit. So we said, "We'll put in $100,000 a month if you can raise $1,000,000 a month for the endowment." And they did. I think they raised $4,000,000.

Right now Uno Park is being built. This will be open space and public gardens where people can grow vegetables. We gave this property to MASS MoCA and then had SHoP architects design the entire park. Then we gave MASS MoCA the money to fund the park and they've agreed to maintain it. In a cool way, this is the first visible activity MASS MoCA has ever done outside of its campus.

Joe Thompson, MASS MoCA, Founding Director from 1987–2020

The average overnighter essentially spends seven times as much as a day-tripper. That means it's far more effective, from an economic and community redevelopment point of view, to convert day-trippers to overnighters rather than to attract new day-trippers.

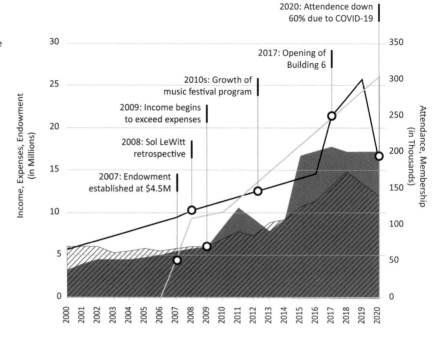

Mass MoCA
Since its opening in 1999, the institution's visitation, income and endowment have steadily risen while keeping expenses in check. *Source: MASS MoCA*

We've grown visitation nicely over the last ten years or so. People have made a commitment to come to the Northern Berkshires. Now we want to just provide a good reason for people to stay longer.

We opened in 1999 with 200,000 square feet renovated and by 2016 we essentially doubled the development footprint. During much of that period, through 2008–2009, we were in survival mode, trying to grow the complex, trying to do good programming. We had no endowment, no cash reserves, no lines of credit. We received virtually no money from our state for operations or programming. We raised private money because Governor Weld, a Republican, insisted. I was initially resentful, but it turned out to be one of the greatest things in the world. It forced us to go out and make friends, develop our members, and essentially sell the project to many, many private supporters. That became the base for our continuing support even after we got the building done.

In 2010 we teamed up with Wilco and did the first Solid Sound Festival. For the first one we had 3,000 people. Last summer we had 10,000 people a day. And it changed the way that people thought of MASS MoCA. All of a sudden we were not just a museum, we were a major venue for large acts. It's a civilized place to see outdoor concerts. You can come and see the museum during the day instead of standing out in some farmer's field getting all sweaty and muddy.

Jack Wadsworth, the owner of the Porches, reasoned that if you love Williams College, then you need to love Williamstown. But if you love Williamstown you need to love North Adams too, because North Adams was the small urban battery that was cheek by jowl with Williamstown and it could be a healthy and vibrant place instead of in economic disarray. He was thinking: *make North Adams a feature for faculty and staff and students.*

Jack was walking down the line of galleries and looked out on these abandoned buildings on River Street and said, "You need to buy those, because it's the one place you look out of MASS MoCA and you see the town. And that looks like crap." I said, "That's a great idea. You should do it. We're loaded down with more real estate than we can possibly digest." About three months later he called from Tokyo and said, "Somebody in the private sector needs to nominate themselves to lose a lot of money in North Adams. And I guess I'm going to nominate myself to be that person."

We did it quietly. I would buy them and he would pay and so all of a sudden we had these seven buildings on River Street. I think it was $350,000 to buy the whole two blocks. We opened the first building in summer 2001 and by the end of the summer of that year we had three or four of them open and then the next year we finished the first round.

As a pure philanthropic donor, Jack has been one of the key people who has nurtured MASS MoCA and kept it alive. We needed to have any form of income that we possibly could, so we pitched him with, "Let us run your construction project and pay us the fee. Why pay somebody else?" We had just built 200,000 square feet—we knew how to hire and fire contractors and had project management skills. And he said, "OK, you guys know how to build."

For three or four years, when we were struggling to stay in business, we took on development projects for a fee that would buy heat and power and pay salaries and keep the museum afloat. We also undertook commercial real estate development on site. We now have 35 commercial tenants on Marshall Street leasing 125,000 square feet of space for about 15 percent of our total budget—about $1.8 million out of $11 million per year.

We're 100 percent occupied now but for many years we percolated along at 50 percent or 60 percent occupancy. The first lease was for $1.25

The refurbished mill buildings of MASS MoCA in the town of North Adams, Massachusetts with Berkshire Mountains beyond.
Credit: Image courtesy of MASS MoCA/Sofia Taylor.

square foot per year in 1997. It was a Hollywood special effects firm that leased 1,000 square feet and paid us $1,500 for a one-year lease. We let them use our coffee machines to close the deal. And now our average lease rate is more like $14 per square foot.

We team up with other schools, universities, foundations, artists and make a grant trade. The trade is, we have space, time, a talented staff and a growing audience interested in new ideas. Let us host your content. There are urban institutions that have stuff but real estate is expensive and they don't have room. We take on all the tasks of opening, closing, security, education, taking tickets. We run the place and they get to show their stuff. Probably a third of our total footprint is art that comes through those kinds of long-term, programmatic link ups. Some are fifteen years or twenty-five years. Some came with money but for some we had to raise money. It's not a one-size-fits-all solution. It's like a portfolio of relationships in some ways.

Nadine Berger, Former Associate Principal, SHoP Architects (until 2022), Sustainability Senior Manager, AECOM

People come to MASS MoCA, stay there for two and a half hours, and then they move on, but nobody really goes downtown after visiting Mass MoCA. In the last few years, they've tried to change that. They added signage "two minutes to downtown" and they've really made an effort to connect and to acknowledge that they should do more to help the town. That's been great to see.

In 2013, we showed them that they had a lot of asphalt. Urban renewal razed the south side of downtown and destroyed the fabric. The city has so many parking lots and non-porous areas, but for green space, common space and parks, there's nothing there. It was really a sad scene. Jack Wadsworth, the owner of Porches, wanted to show that you can add green space, so we worked with him on a mini-project to create Uno Park. There was green space around it, a kind of oval, and a little playground. We expanded that and opened it up to the street to make it visible for the community.

The big thing that we suggested was to have a Town Green in downtown, just off Main Street where you have the big Shoppers Square. We called it the Commons and suggested a big green space with a carousel and a field where they can do ice skating in the winter and a green market—a communal space in the center of town that combines the idea of enhancing the quality of life for residents and creating an attraction for tourists.

The second thing is creating this link between downtown and MASS MoCA. In our study, we put sculptures everywhere. The idea was to have these markers that also expanded Mass MoCA's footprint outside of their campus. There was a big sculpture at the entrance to the park, and another at the corner of Marshall and Main Street. That one—Big Bling—was installed last year at the link node. You see it from downtown and you see it from Mass MoCA. As a landmark at this junction it gets people around the corner to downtown and hopefully into the stores.

View from the highway underpass of Martin Puryear's temporary (five-year) sculpture entitled "Big Bling." The 40-foot-tall wood sculpture on the periphery of the museum is adjacent to the central business district. Its strategic-location on the other-side of the highway draws visitors from the museum into downtown.

Simeon Bruner, Architect and Founder, Bruner/Cott Architects

We had three goals: to create a brand-new art museum, to preserve an historic mill complex, and to become a catalyst for downtown North Adams. It started in 1987. There was an RFP [request for proposal] that the state put out, pushed by Tom Krens, for a study. We wound up on a team where the lead architects were Frank Gehry and Bob Venturi. SOM was the prime contractor. We were the mill experts on that team. The state did a study that proposed a $35 million grant to rebuild the place. But the proposal was basically to make a Club Med. It was going to use art as the decoration within the complex with a lot of condos. The state funded the bond issue but the governor wasn't going to fund the development.

Apertures filter natural light through layers of facades and double height spaces at the MASS MoCA complex that includes more than two dozen buildings. Credit: (left) © Michael Moran/OTTO; (right) Authors.

Then, in the late 1980s, Joe Thompson took over. He wanted to find a way to make it go. And his original idea was to take the second floor. This is like a three-quarters of a million square foot complex! In my view, you rethink the program to fit into the building. But SOM had never done any rehab. Every time they looked at a building they said we can change the windows, or it only costs another $10 per square foot to put in new windows and heating in, and now that we've got heating why don't we put in air-conditioning and repaint the brick. It went from a straightforward approach to trying to retrofit a building to do something that it was never capable of, as opposed to exploiting what was in the nature of the building itself. Then Thompson hired us to do a new master plan in phases that would exploit the buildings and be much more affordable.

We came up with three principles from the design perspective. Stabilize the entire complex and do the buildings in the worst shape first. Preserve the buildings on the exterior of the complex to be used for commercial use, essentially as endowment. And push the entrance of the museum as far into the complex as we could to move people through.

In the first phase we dealt with the worst building, which was Building 4. The floors were rotten. That's why you have all those three-story spaces—where a floor is rotten it's cheaper to just not put it back. We exploited the north light and we put back only the ground floor. You exploit what's there to make a space that you might not otherwise create. We had no money, but we had tons of space. We asked, "How can we use all of this space to create something that's beautiful?" So we created a sequence of linked galleries where there's no corridor and light is used to show you around.

You can't find mill buildings that are affordable in towns that have an economic base anymore because they're too desirable. But if you back up and ask yourself, what do you have as an asset? Then you're getting down to the fundamentals. We found our river in North Adams because the mill was there and then we put that together with the art.

Ben Svenson, Broder Development, Developer of Tourists Motel, Entrepreneur

When we started to come here, I met John Stirratt who was my partner on this project. John wanted to open up a small hotel and although North Adams wasn't necessarily on his radar, he knew it well because his band Wilco has a music festival here.

One of the things that really hooked us was this idea of early tourism in North Adams and that people came here a hundred years ago in droves. There were hundreds of accommodation businesses along Route 2. But in 1914 they completed the modern Mohawk Trail right when people were starting to get cars. In the 1910s you had over 100 cars an hour doing the hairpin turn. People would start in Boston and make the journey west.

Skip forward a long time, to MASS MoCA's creation and people started coming here for cultural reasons to go to the museum. And I came to the Museum ten years ago with my wife and fell in love with MASS MoCA. I'm a buildings person. The buildings, at times, are more interesting than the art.

There's a lot of old motels buried within this community. I saw a "for sale by owner" sign, called up the owner, saw the rooms. They were terrifying; sticky floors, broken everything. I put on my snowshoes, walked behind the building and experienced this very unique condition where, because we are in a flood plain, the earth drops off about 15 feet and you just see and hear the river and lose your sense of the built environment. The buildings, the road, all that goes away and you're in this Berkshire bucolic wonder-moment.

Every room had a six-foot by four-foot picture window looking at the road and on the backside just had a little vent window in the bathroom. Maybe that made sense in 1962, but in 2018? Today the river is interesting and no one wants the road to be emphasized. We wound up taking down all of the original motel though we reused some of the foundation and the iconic facade and shape and massing. We put all of the glass in the back, but we wanted people to know when they came here that there was something before.
. . .

Pared down but pricey, the Tourists is halfway between North Adams and Williamstown. The boutique, forty-eight-room hotel is a former 1960s roadside motel re-built and cladded in sleek white oak, demonstrating small town contemporaneity and shifting orientations from road to landscape. Room rates in 2024 were near $400/night with minimum two-night stays in high season.

I loved all of the Lawrence Halprin "Rules of Engagement" and William Turnbull Jr. was my favorite of that group of architects. Natural materials, untreated on the exterior. They used redwood; what could we use? How about some clear white oak? There sure are a lot of barns in New England made out of that. Make the outdoors the artwork and make the insides quiet, like a portal to view what's outside. Don't compete with nature.

I wanted to build a really good building. A Days Inn or Hampton Garden Inn is just a kit. They'll have a lower rate, and they'll make more money too, by the way. Our building took a lot of planning and soft energy and we built it to last eons. The white oak siding will survive you and me and our children. This is a LEED platinum building and is as good as you can build in 2018. The unfortunate reality is that in order to do that, you have to charge a rate that is the most expensive hotel in town.

John Barrett (D), Mayor of North Adams, MA from 1984 to 2009; Member, Massachusetts House of Representatives (since 2017)

When I came into office in January of 1984 the largest employer in the city, Sprague Electric, announced that they were moving their corporate headquarters. The President of the company says they're going to Worcester because, as he put it, "North Adams is a dying, depressing area."

That's when I brought in Governor Dukakis and said "What can you do to help us?" We spent almost two years trying to figure it out. Let's make Sprague Electric into a prison, a fish market, you name it.

I had been a county commissioner, so I understood things and I knew some people. The first thing was to change the image of the city. So we borrowed money to fix the streets and paint the fire hydrants and cut the grass. We were going to rip down the old houses that were decaying, clean the city up, and couple that with the renovation of the old railroad yard that Dukakis had initiated prior to my getting there. In order to attract you have to be attractive to the outside world. It was a gimmick to get people believing it was going to be OK. Because then, when MASS MoCA came along we had laid the groundwork that this project wasn't for elitists. In blue-collar communities there's always this class warfare that you have to deal with, but I could say, "No, we're going to benefit from this."

It took three Governors and 14 years to the day from when we announced MASS MoCA to the day that it opened. It started with Dukakis who had provided a couple hundred thousand to do the planning. They threw him out over the recession that hit in the late 1980s. The first thing Weld did when he got in was to cut that money. Williams College had put in a lot of money, so I called up their President, Oakley, and said, "I've got an economy over here that's dying, and you're sending kids all over the globe with your professors to fix these third world economies. Do you think you could maybe send them a few miles down the road?" Because they had been an isolated institution living in their own world. And that's why they stepped into the picture.

Prior to my getting here the south side of town had been torn down and replaced with a strip mall and a Kmart. The most critical buildings were lost: A hotel with marble staircases, and banks, for instance. I determined that I had to run the city like a benevolent dictator to get things changed, while at the same time including people to make sure we got their ideas. And I had to find a way to use MASS MoCA as a catalyst to get things done.

Outsiders are now dictating what's happening in North Adams. They only care about making a buck and about what's going to benefit MASS MoCA. And you've seen destruction, like what's happened to Heritage Park. I had it filled. The city had to take it over because a developer couldn't make it work. We don't need more art. We need to create a business center.

I wouldn't let the Social Security office move to MASS MoCA. They said "Why? We could get all their rent money." And I said, because you're looking only at your situation. I have to look at taking them off of Main Street. As soon as I'm out the door, I bet it was within six months they were over at MASS MoCA and off of Main Street. And then MASS MoCA continued to take large law firms off of the Main Street. It's sucked the life out of downtown.

A Conversation

Ana Farrington (AF), Gallery Owner

Emilee Yawn (EY), "The Plant Connector," Shop Owner

EY: MASS MoCA has a gate around it, like an ivory tower. I love MASS MoCA and I'm a member and I came here because of MASS MoCA. But there's such a division between the institution and the city.
AF: I have a slightly different perspective. I also came here because MASS MoCA is here and real estate is relatively inexpensive. I bought a few years ago and I had this opportunity to open a gallery and envision something I never could have had in Cambridge, Massachusetts. What I see is a lot of people invested in making that connection between the museum and the city and the college. Because the college too is a big player that people underestimate as a generator of tourism and new vitality.
EY: North Adams definitely doesn't have that college town feel to it. You don't see young people out and about. To me, the biggest highlight was when college kids were excited to be in my space. Yes, I got MCLA [Mass College of Liberal Arts] kids here!
AF: I think one of the big problems downtown right now is the large number of properties being held by owners who live elsewhere. This large brick building right on Eagle Street was bought by the same person that owns the giant building right there. He lives in New York and hasn't developed either of these properties, and one of them has fallen into such disrepair that commercial spaces on the first floor can no longer be rented. But the city doesn't have any way to leverage pressure on this landlord to encourage him to redevelop his properties. There's no vacancy tax in the commercial business zone, no fee or fine, and the buildings are just crumbling.
EY: I'm from San Francisco and I moved here because of the Assets for Artists program, MASS MoCA's initiative that brings artists to North Adams. I feel that what Ana and I are doing are small moves, and grassroots. But still everyone's like, "You can't have a price point that's a decent amount because this is

View down Eagle Street in downtown North Adams, Massachusetts.

a very depressed area." The truth is people want good things and they don't necessarily want to just go to Walmart. They want something different, and innovative. I think my installation space is a good example of innovation because it changes all the time. The street is always different because the installations are always changing.

AF: I do get a fair number of visitors from MASS MoCA. However, I've also found that by the time the MASS MoCA people get here they're like, "We can't look at any more art." People have literally said that to me.

EY: To open a business is really quite hard. There are a lot of hoops you have to go through with the city. This was a record store before, meaning I had to do a change of use. Anytime a building changes its name it's a change of use. But that doesn't make sense when it goes from being a retail store to being a retail store. The planning board only meets every six weeks and then you have to wait twenty-two days for anyone to contest it. After that you have to get an inspection from the permitting department. And then you have to go and get your deed. It was three months of fronting rent before I could even open up my door. But I don't think I could do what I'm doing anywhere else. There's a community here. People who transplant here are so pioneering in their mentality.

AF: Right. It's like a Wild West mentality of going out to the frontier. It's putting a stake in the ground to do your thing.

Notes

1. 2021 population (U.S. Census Bureau).
2. Massachusetts law states that no town with a population of less than 12,000 may adopt a city form of government and no town with a population of less than 6,000 may adopt a representative town meeting form of government.
3. Established by Congress in 1991, there are 109 National Scenic Byways recognized by the U.S. Department of Transportation.
4. The original museum was designed by Daniel Perry and opened in 1955. The 1973 Manton Research Center was designed by Pietro Bellusci and The Architects Collaborative. A 2001 Master Plan for the campus was completed by Cooper, Robertson and Partners and there are two Tadao Ando buildings. The landscape design was by Reed Hilderbrand.
5. The Five Colleges are Amherst College, Hampshire College, Mount Holyoke College, Smith College, and the University of Massachusetts Amherst.
6. MASS MoCA is a National Register-listed historic site, and so therefore any demolition of the complex had to be justified on historic grounds in addition to other criteria.

References

Berger, N. (2022, August). D. Gamble and P. Heyda, interviewers.
Sprague, J.L. (2016). *87 Marshall Street, Creation, Disruption and Renewal in the Northern Berkshires*. Williamstown: MASS MoCA Publications.
Wadsworth, J. (2018, June). D. Gamble, interviewer.

1.2
Jamestown, NY

Jamestown (population 28,393)[1] is in the far western corner of New York nestled between Lake Erie and the Allegheny National Forest. Located at the end of Lake Chautauqua, it served for generations as fishing and hunting grounds for Native American Seneca and Erie people. Later this region's pastoral geography and abundant fresh water was subsumed into a leisure and manufacturing center, spurred by the rail and waterways that linked it to larger nearby population centers like Cleveland, Pittsburgh, Buffalo and Erie.[2] The story of Jamestown today is bracketed by this 19th century industrial growth, its gradual decline in the 20th century following deindustrialization and the departure of the train, and its quest today for stabilization with a resurgence in tourism, entertainment and culture.

Similar to North Adams, Massachusetts, Jamestown's economic model relies on attractions downtown and drawing on assets in neighboring towns, while maximizing its own position in a bucolic region that is itself a destination. The most influential of Jamestown's neighbors is the village of Chautauqua less than 20 miles (32km) away along the Chadakoin River and across Chautauqua Lake. Home to the renowned Chautauqua Institution, since 1874,

Late 19th and early 20th century mills and manufacturing align along river and rail, spurring downtown growth but snubbing nature. View of Jamestown, New York looking northwest toward Chautauqua.
Credit: Fenton Historical Center.

an "experiment in educational and vocational learning," the village attracts as many as 100,000 people every year in the summer for nine weeks of lectures, music, and art. Jamestown's contemporary experiment builds similarly from the idea of scripting place by elevating programmed experiences. It has become the nation's heart of comedy and performance, concentrated at the former train depot and riverfront from which the historic town began.

Like many former manufacturing centers, Jamestown's physical legacy left mills and major swaths of rail infrastructure along the river edge, in addition to handsome but abandoned buildings like the train depot. As many as 120 factories and mills obscured the once pastoral Chadakoin River that passed along the rail line through downtown. Predictably, with finite local resources and a declining manufacturing base, steady and significant population losses ensued as Jamestown, like North Adams, struggled to diversify its economy towards the end of the 20th century. Aging neighborhoods, derelict infrastructure and a sparsely populated business district fostered high poverty rates that coalesced around a languishing core. Today, Jamestown is smaller than at any time since Theodore Roosevelt was President, but it occupies—and importantly must pay to maintain—costly infrastructure built for a much larger population (czb, 2010).

Jamestown has slowly staged change, by changing the stage. One of the first moves toward drawing new investment to the rundown core was to improve the outward appearance of the substantial inventory of historic buildings. An urban design plan process led by Boston-based Goody Clancy was initiated in 2006 to address Jamestown's shortcomings through several parallel initiatives. Funded by the town's largest philanthropic organization, the Gebbie Foundation, the facade improvement program was all about appearances—a strategy that literally propped up downtown as a stage set. Buildings were made to look healthy even as underlying economics and markets still needed support. But the physical improvements slowly cultivated investment confidence. Other initiatives followed that could then shore up realities behind the facades. These were led by both public and non-profit actors and revolved around securing major state and federal funding for infrastructure and public realm improvements more broadly.

A five-term mayor (with a background in urban planning and economic development) provided longitudinal consistency in the revitalization process that spanned decades. But if the time of transformation was vast, its geographic concentration was tight. Building revitalization efforts focused on one of five sub-districts downtown that was just one-half-mile-square, or 14 city blocks. The plan returned investment to the downtown core and rail and river corridors adjacent to it. This strategy enabled a focus on the defining features of Jamestown, its waterfront and the historic buildings—but also on its historic legacy in comedy. Jamestown is a funny place: it is where comedienne Lucille Ball was born, and where the National Comedy Center now calls home.

Recentering Civic Life

In many older American towns, the railroad station or depot was the most important building and center of community life. Stations were places of exchange, commerce and communication. Depending on the number of companies that served them, even small towns could have three or four train depots for connecting residents to work and leisure in the larger metros or region (Halberstadt, 2002).[3] But unlike a civic town hall or post office, depots were not technically "public." They were privately financed and constructed, designed by enterprises in direct competition with one another. The architecture of the station was integral to the line, and stations functioned as brands and advertisements for the companies that built and used them. Jamestown's main depot was designed to uphold the town's highest civic aspirations. It was located at the base of a hill, teetering on the shift in topography adjacent to the river at West Second and Washington Streets.

Nearly fifty years after the railroad first came to Jamestown, its rail infrastructure was fortified and thickened. In light of the many rail crossings within the tight street grid, the city constructed a railroad viaduct during the 1920s. The new rail elevation separated it from the streets below, with North Main Street, Foot Avenue and West 2nd Street passing below the tracks. The viaduct facilitated automobile circulation and removed conflicts between the rail line and the roads, but its construction reinforced an edge not previously as prominent. **Downtown became severed from the river**.

The prominent railroad viaduct where it crosses North Main Street, simultaneously stitching and severing.

1.2 Jamestown, NY

Following the construction of the viaduct, a new Art Deco-style depot was constructed in 1932 for the Erie Railroad, replacing the older 1897 Victorian-era structure nearby. Modern and streamlined, the new depot spanned three stories between 2nd Street and the tracks as a result of the viaduct construction. Sadly, the building proved to be a monument to an industry which had already seen its peak, as passenger traffic fell precipitously following World War II. By the time the last passenger train rolled through Jamestown in 1970, the station was severely neglected. Lacking its primary purpose, the building was vandalized over time and poised for demolition.

With mounting community concern about the fate of the historic building, the Jamestown Redevelopment Corporation (JRC) acquired it in 1993 with $120,000 of Community Development Block Grant (CDBG) funds. The historic train station was stabilized, cleaned and preserved. It was listed on the State and National Historic Registers and awaited new life. Occupying a key location downtown, the property nevertheless remained vacant until a viable reuse strategy would emerge. The National Comedy Center eventually gained traction as a possibility and the facility opened its doors in 2018. The center of local civic life would now be focused on *tourists*, rather than workers and goods. While this is a common revenue-generating strategy for cities and towns alike, it signals a subtle shift in priorities from civic space as fostering local life, to fostering visitorship. There are more reasons why cities and towns have undergone this transformation, and they are related to broader manufacturing shifts away from towns, coupled with severe government funding cuts to cities and towns since the 1970s.

Funny Business

Still, the tourist draw in Jamestown is rooted in the town's genuine pride in the people who grew up there and have a strong connection to the place. For such a small town, Jamestown is fortunate to lay claim to three influential figures who were born or initially resided in the area. Artist, educator and naturalist Roger Tory Peterson (1908–1996) has a Robert A.M. Stern-designed institute dedicated to him. U.S. Supreme Court Justice Robert H. Jackson (1892–1954) is renowned, among other things, for serving as the chief counsel at the International Military Tribunal for the prosecution of Nazi war criminals. His center downtown is in a renovated mansion. But it is the legendary comedienne Lucille Ball (1911–1989) whose larger-than-life face graces building facades and whose memory grounds the town's identity—and a portion of its economic development strategy.[4] The Lucille Ball-Desi-Arnaz Center was established in 1996 in a renovated two-story building at the intersection of Main and Second Streets to celebrate her legacy, and the Lucille Ball Comedy Festival brings in thousands of visitors downtown every summer. The success of that endeavor spawned an ambition to grow Jamestown's role as center of comedy beyond Lucy herself and grew into the National Comedy Center, an immersive and highly interactive national repository of comedy located in the renovated former train depot and an adjacent new building. Jamestown's rebuilding literally pitted itself as the center of a national pun—the home of comedy in the United States.

The critique of tourism as economic development is that it focuses less on the existing local populations and more on drawing people from the

Many Jamestown buildings form the laughing stock of the town, with larger-than-life murals of Lucille Ball and iconic scenes from her legendary *I Love Lucy* show.

outside. Larger cities have long relied on destination-type investments to draw visitors and generate sales tax and other revenue (Swanstrom and Judd, 2006). Some of the most famous American models of this are the festival marketplaces and aquariums that can be found at Boston, Massachusetts and Baltimore, Maryland's harbors. This tourist-centered urbanism is often referred to as the "Rousification" of cities (after developer James Rouse who championed the formula in the 1970s and 1980s). Fittingly, they have been critiqued as a "deliberate creation of the city-as-stage" (Hall, 2002), reproducing aesthetic pockets of sanitized urbanism to foster consumer culture that masks other areas still lacking basic resources and services.

In this context, Jamestown's effort to secure the National Comedy Center is part of a predictable revitalization pattern. But for a small town, it represents a noteworthy twist on the paradigm. The scale of impact of a national destination-anchor arguably goes a long way in a town of fewer than 30,000 people. Plus, the comedy center is based on extending the legacy of the town's most treasured resident. The initial impacts of the center, alongside the urban design plan and facade improvement efforts, have led to major federal and state funding for core infrastructure. Those public investments eventually spur new infill development.

Plus, as larger cities with destination-attractions compete with one another, smaller towns like Jamestown—with its neighboring Chautauqua Institution and easy access to the Allegheny foothills and Lake Erie —increasingly share an ethos of *regional collaboration*. **There is a shared recognition that the town destination tourism model works better with more settlements and attractions participating, not competing.**

1.2 Jamestown, NY

A 21st century anchor for the 19th century downtown, the National Comedy Center occupies the rehabilitated Erie Train Station (rear part of the building in foreground) along with new construction.
Credit: National Comedy Center.

The most successful museums are entertaining and fun, with immersion experiences that make the past more relevant and contemporary in the present. In this regard, the National Comedy Center—the only museum dedicated to the art of comedy in the country—is at the forefront of interactive engagement. Its approach to education is highly digital and *customized*, with electronic exhibits that collect data to tailor visitors' experiences in real time. Upon entering, participants spend time at a kiosk to select comedians, TV shows and movies that they find funny. An electronic bracelet loads this information and visitors get matched with things that makes them laugh as one progresses through the exhibits. The programming gets more and more specific to a visitor's tastes as information gets incrementally added from the many exhibitions. This strategy ensures people return, since their experience will vary each time.

In 1937 urban theorist Lewis Mumford wrote about the vibrant city as a "theater of social action," where people are both spectators and participants in the scene of a bustling city (Mumford, 1937). Does the comedy center represent an ultimate privatization of this theater, as a crafted urban destination experience (now digitally tuned to even capture the unpredictability of the real city?) Or, does the comedy center's presence enable real acts of urban life to play out once again around the civic train depot?

Comedy can be profane, of course, and access to the "Blue Room" of the comedy center is restricted to guests aged eighteen and over who are not easily offended by raunchy humor. The space is connected by an elevator to the lower (riverside track) level, which takes advantage of the change in grade between the building's front at the street-level, and a new park at the lower, waterfront level. The high-tech attraction came with total price tag of $40 million, requiring a complex layering of public financial support including State and Federal Historic Tax Credits, Community Development Block Grant funds,

The riverfront park area that was once lined with train tracks behind the historic station in Jamestown, New York (at right). That station is now the National Comedy Center (NCC) that includes lower-level exhibits accessible by elevator and ramp connections to an outdoor terrace overlooking the park and river.

and contributions from the Federal Transportation Authority, Empire State Development Corporation, the State Department of Transportation and New York State's Environmental Protection Fund. The park and riverfront vision that the 2006 urban design plan initially forged may well be the most important part of the story of the town as theater, as those efforts steered investment towards crucial connective infrastructural fabrics that ultimately serve both local residents and international visitors.

Keeping Up Appearances

The bulk of Jamestown's building stock was constructed in the late 1800s and early 1900s, so the age of much of the downtown fabric has easily passed the century marker. Fortunately, many of the buildings with significant architectural character remain intact. There are large churches, institutions and a wide variety of residential typologies in close, walkable proximity to each other. The ensemble of typologies makes the downtown appear like a model railroad only *full scale*. Just outside the core, however, many residential neighborhoods suffer from neglect and deferred maintenance with more than three-quarters of the city's housing stock built prior to 1940 (czb, 2010). Though the pace of population decline has slowed, the town continues to lose residents.

1.2 Jamestown, NY

Vacant buildings, regardless of how remarkable their architecture, are vulnerable to decay and demolition. As identified in the *Economics of Historic Preservation*, when a building is empty and unused, four things can be done: (1) nothing; (2) stabilize the building and wait; (3) rehabilitate the building; or (4) tear it down. Option 4 is always available, but if that option is taken all the others are eliminated (Rypkema, 2014). In Jamestown, as in North Adams, Massachusetts in this book, each of these stages materialized to varying degrees, including the fourth phase when a considerable swath of the downtown on the opposite side of the river was removed in the 1970s during urban renewal.[5] Still, within the undemolished core, Jamestown's transformation started at the surface-level with a facade restoration program. If the extent of the building stock couldn't be fully renovated, it could at least present a new face of improvement. The improvements were effective, concentrated around the intersection of Main and 3rd Streets.

To further revitalization efforts, the downtown core was listed on the National Register of Historic Districts in 2014 with over 100 contributing properties. The designation enabled up to 20 percent federal income tax credit for substantial rehabilitation of historic, income-producing properties, and 20 percent state historic tax credits to be used in tandem with the federal credits.

Evidence of adaptive reuse and historic building and facade restoration in hilly downtown Jamestown, New York.

Taken together, these subsidies enhance the likelihood of adaptive reuse. In an effort to stem further disinvestment and revitalize the core, the Gebbie Foundation (one of *ten* local, well-endowed foundations and Jamestown's largest) focused their resources on downtown's physical environment. They provided grants through a rebate program to business owners to improve their building facades. One by one (from 2007 to 2010) over thirty-six building elevations were rebuilt and restored. The improved fronts set the stage for further funding and deeper investment, which, in turn, stabilized and rebuilt investment confidence.

Core Planning

Both the National Comedy Center and the facade renovation built off of the 2006 Goody Clancy Urban Design Plan. The plan concentrated formal and organizational design investments on the core and the riverfront, and divided the downtown into five sub-districts that each shared similar physical and land use characteristics. The strategy reinforced 3rd and 2nd Streets as primary east–west connectors and Washington Street (which spans the river as a fly-over bridge) as a primary north–south corridor. The urban implementation was led by the Jamestown Renaissance Corporation.

Within a decade, enough had been accomplished from the first Urban Design Plan that the architecture and planning firm was invited back to Jamestown in 2016 to complete a Master Plan Update. Since 2006, city services had been streamlined and the town's bond rating improved. In the residential neighborhoods, the Chautauqua County Land Bank Corporation partnered with the city to remove forty "blighted" properties. Enhanced partnerships between the local and state government officials,

New pedestrian bridges and an expanded Riverwalk along the Chadakoin River in Jamestown, New York.

not-for-profit-sector and private investors were beginning to change the look and feel of the Central Business District (CBD) and the Greater Jamestown Riverwalk Trail was emerging at the foot of the former train depot. A $1.6 million Transportation Alternatives Program grant, together with a $400,000 local match, enabled two pedestrian bridges to be added over the Chadakoin River, connecting residents in the south and west-side neighborhoods with the Riverwalk to employment centers in the CBD. These bridges, and the riverwalk trail system, signal investments now more than just skin deep. **The town has been able to move from the stage of facade stabilizations to crucial urban circulatory systems, something that enables access across the city for locals.**

The urban design update coincided nicely with Governor Andrew Cuomo's New York State economic development program. The Downtown Revitalization Initiative, or DRI, was created to stabilize downtowns across New York through the adaptive reuse of existing buildings to address job creation, enhance connectivity and promote public transit and "smart growth" principles.[6] It was anticipated that the majority of funds would be directed towards capital projects in order to transform a downtown's physical environment.[7]

In 2016, Jamestown was selected in the first round of that DRI funding for the Western New York Region, bringing in $10 million and providing a significant boost to the city. The monies were dispersed across numerous initiatives including the private redevelopment of catalytic sites and upgrades to tourist destinations, and improvements to the Riverwalk and downtown programming for residents. The Master Plan Update was completed in 2019 and advanced the initial urban design goals of making the downtown more walkable, including the renewal of the Chadakoin Riverfront as a community jewel. It also capitalized on an emerging health and wellness district south of the river, in the less-dense area once called Brooklyn Square that was destroyed during urban renewal.

Region Relations

There is no magic formula between the size of a city and the scale of its attractions. Other New York small towns also have disproportionately large tourist attractions. Cooperstown, for example, with a 2021 population of just 1,867, is located 60 miles (97 km) southwest of Albany and home to the National Baseball Hall of Fame and Museum which attracts nearly 300,000 per year. Along the southern tier of the state, Corning (population 10,738 in 2021) is a four-hour drive from New York City. It has an internationally renowned Museum of Glass which brings over 400,000 visitors per year. Like North Adams, Massachusetts, these two New York communities offer more than just a single tourist venue, aggregating multiple experiences together *between towns* to keep people in the region longer.

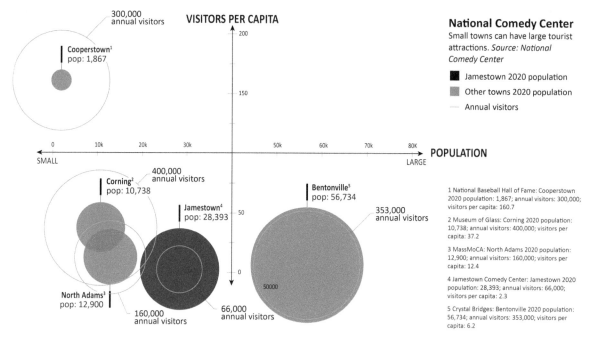

National Comedy Center
Small towns can have large tourist attractions. *Source: National Comedy Center*

1 National Baseball Hall of Fame: Cooperstown 2020 population: 1,867; annual visitors: 300,000; visitors per capita: 160.7

2 Museum of Glass: Corning 2020 population: 10,738; annual visitors: 400,000; visitors per capita: 37.2

3 MassMoCA: North Adams 2020 population: 12,900; annual visitors: 160,000; visitors per capita: 12.4

4 Jamestown Comedy Center: Jamestown 2020 population: 28,393; annual visitors: 66,000; visitors per capita: 2.3

5 Crystal Bridges: Bentonville 2020 population: 56,734; annual visitors: 353,000; visitors per capita: 6.2

There are easy critiques of urban redevelopment so overly focused on people who live outside of town. This is especially true in racialized weak-market economies where locals remain without investments in underlying basic services while tourist economies geared toward visitors are propped up. Jamestown's model is noteworthy in the *regional* context, as infrastructure investments elevate local experiences and build connectivity with the larger region that in turn relies upon the downtown as a hub for commerce and culture. The scales intertwine.

Jamestown's position resides within a regional tourist-ecosystem of small towns and attractions. The American urban redevelopment paradigm that targets revenues from tourist activities stemming from a main attraction is broken into a constellation of attractions where towns and villages are concerned. As with the cooperation between North Adams, Massachusetts and its MASS MoCA (Massachusetts Museum of Contemporary Art) and nearby Williamstown, Massachusetts with Williams College and the adjacent Clark Art Institute and other nearby venues and towns, Jamestown and the Chautauqua Institution across the lake have strengthened their programming partnerships. A century ago, access was easy: large steamer ships and trolleys departed directly from Jamestown's downtown railroad depot for Chautauqua Village. Unfortunately today, beyond the lake tours, actual regional connective transit between the two towns has been slow to materialize.

The reality is that many American towns struggle with isolation because the historic connections that facilitated their development patterns

The restored steamer *Chautauqua Belle* heads towards the Chautauqua Institution's Miller Bell Tower and Pier Building. Credit: Courtesy of Chautauqua Institution.

no longer exist, or have been substantially eroded. When one segment of transit is severed the entire network deteriorates, leaving the car as the only means of (unsustainable) connection. Nevertheless, long-term ambitions for a regional train connection in Jamestown persist, including one that would link to Buffalo, New York 75 miles (121 km) northeast and on to Niagara Falls some 20 miles further (32 km). Located at the border between New York State and Ontario Canada, Niagara Falls (tagline: "where adventure falls naturally") attracts between eight and thirteen million visitors annually, making it one of the most popular international tourist destinations in the United States. Such an impressive vision would reinforce and revive the aging railroad network at scale. Growing the rail network regionally—as well as the local pedestrian and bike trails—is a crucial next step.

Perspectives

Howard Zemsky, Former Executive Director, Empire State Development Corporation

Jamestown is an hour and a half from Buffalo but it's also pretty darn close to Cleveland, Pittsburgh and Erie. Upstate New York isn't that far from Downstate New York. So you've got dense population centers, depending on the radius of the circle you draw around these places, and they can pull from a wide area. Upstate, it doesn't matter if you're talking about Jamestown, Buffalo, Binghamton, Syracuse or Rochester; most every city struggles with the same thing: A long history of disinvestment and sprawl to the suburbs. We've lost population Upstate, and spread out over twice the land mass.

Many decades of neglect downtown created a higher cost environment but also a less vital one. So, the Downtown Revitalization Initiative (DRI) is an attempt to reverse those effects and put a policy in place that will help revive the core and keep and attract a skilled workforce.

Whether you call it smart growth or revitalization, we're always looking to invest in the core. To activate the waterfronts. To repurpose the historic buildings, like old manufacturing facilities that can become mixed-use commercial/residential, maybe retail. Or hotels that can be brought back to be amazing places to stay and have a rich history and unique character. The character of these downtowns to me are defined in part by the potential reuse of their buildings. They're beautiful. There are too many areas around the country where the stock of historic buildings were lost.

Part of the Comedy Center is the rehabilitation of an old train station and it checked our boxes in a lot of ways because it was historic preservation, adaptive reuse, waterfront activation, tourism, and it leveraged the proud history of Jamestown and Lucille Ball. And it had something else that was important—support from the community. **You have got to create a density and vitality downtown, and where you have big, vacant structures in strategic locations the community has to rally together and figure out how to repurpose and revitalize those.**

All of these projects add to our *ecosystem of tourism*. We want people to stay longer, to take in Niagara Falls and the Darwin Martin House and the Albright Knox Museum and go to the National Comedy Center and to the wineries, whether in Western New York or the Niagara Wine Trail or Chautauqua Wine Trail or Finger Lakes Wine Trail. They all build on one another.

I often say we've gone through forty years of decline and you don't bring it back in five years. To some extent it's having the right plan for what you're trying to accomplish, and it's having the state, in particular, because they're always the 10,000-pound gorilla in the room when it comes to support. We have that now, and the local community actively is engaged and helping to

The historic Reg Theater building in downtown Jamestown, New York.

develop these types of plans through political and civic leadership. So Jamestown is a heck of a lot better than it was ten years ago and more importantly it is moving in the right direction.

David Spillane, Principal, Goody Clancy Architecture, Planning and Preservation Firm, Master Planners

Jamestown has been an interesting story in terms of the dramatic progress that they have made because of the quality of the partnerships between the foundations and the city. They also had a strong belief in the value of planning. It began with Mayor Teresi. Before he became mayor, he was the community development director. So he's a *planner*, you know? In the first interview I had with him all he wanted to talk about was Jane Jacobs.

We were asked in 2007 to create what they called then an Urban Design Plan. The City was rather unusual for a community of this size in that they had an urban designer on staff. He was very passionate about the urban design quality of the city which was very much in decline with all of the usual issues. What was notable about Jamestown at the time as we were interviewed was the participation of several foundations. The Gebbie Foundation was remarkable. They are based in Jamestown and had largely focused on social services before shifting their mission to urban redevelopment.

The idea of the urban design plan at the time was to try and get everyone on the same page. So we started that process and I think it was the most engaged public process the city had ever initiated. We rented a downtown storefront. We did walking tours with people downtown. We had meetings with everybody and people were pretty passionate about the future. We also lived with the parallel track which was talk radio. The talk radio narrative was that nothing was going to work and nothing succeeds in Jamestown, we're wasting more money and it's never going to happen. We went through a two-year process and created the urban design plan and finished it in 2008, rolling into not the brightest of economic moments. In some ways it was a simple plan; a focus on the riverfront which had been entirely neglected. There was a focus on some of the key downtown Main Streets. There was an emphasis on design quality actually being important, that just getting anything built wasn't going to be good enough. All of this was tied to design guidelines that supported the plan.

After we finished developing the plan I said to a client that part of the implementation strategy should be for it to win an award. And it did the next year. It won the Upstate NY APA (American Planning Association) award. I can't tell you the level of excitement that generated. I don't think Jamestown had ever won an award for anything like this. The mayor, the director of community development and the city's urban designer rode with the photographer to pick up the award in Binghamton and it was front page news in Jamestown. It was the first step in creating some kind of external validation for what the community had actually done. And subsequently they founded the Jamestown Renaissance Corporation which was a 501c3 whose core purpose was to implement the urban design plan.

Over several years they adhered closely to the plan, built little pieces by getting $200,000 here and $100,000 there. They built a mixed-use building right next to the ice arena in 2008–2009 on the location of a surface parking lot. They did facade renovations on over thirty buildings on the primary downtown Main Streets. This was really driven by the Gebbie Foundation. They approached building owners and said "Do we have a deal for you." There was a tremendous amount of change there. They did a bunch of work on the downtown alleys and a whole series of things on the public environment that really made a difference. Gebbie made an investment in several downtown buildings. With Gebbie money and some other money they started building a Riverwalk. They got a grant to build a couple of pedestrian bridges that were a part of that. The river was the core. The plan essentially said open up

View down 3rd Street in Jamestown, New York at sunset.
Credit: Courtesy of the Chautauqua County Visitors Bureau.

the river and over time build connections. This connects out to Chautauqua Lake, the summer paradise. Chautauqua Institution is there.

So the core of the plan was strengthening the downtown core as the urban place of the region, trying to connect it to the lake physically and economically. They are chugging along, building a riverwalk in sections and have almost reached the lake. They had an abandoned train station downtown that has been entirely renovated. Large portions of this riverfront that was scrubland has been cleared. Now the train station opens up onto parkland which runs across the train tracks down to the river. We had said in the plan that they needed some kind of attraction that we couldn't define at the time but targeted something that would attract 100,000 visitors a year. Our idea was we create an attraction and pull people in and bring them up into the core of the city. The program was yet to be defined and that has become the National Comedy Center.

After more than ten years of coordinated and sustained efforts Jamestown built an incredible amount of credibility in the region with the state and funders and their ability to make things happen in pretty remarkable ways under difficult (economic) circumstances. About this time there was an initiative out there that the governor announced which was a statewide downtown revitalization initiative. They got the $10 million from this initiative that has continued to support downtown efforts.

I remember thinking at the time, and history has borne this out, that for a community of 28,000 people Jamestown's strong civic leadership was its best asset … there were a lot of communities I was working in that had populations of 250,000 and if I had been able to pick up the Jamestown leadership across the sectors and move them to some of those much larger cities, that would be a more powerful thing than all the plans. For communities like Jamestown, planning is incredibly important and its precision is incredibly important. When you're working in Cambridge, Massachusetts, I mean, things are going to be all right no matter what you do. Here it really has to depend on incredible levels of efficiency and each initiative working with everything else. Everything being networked with everything else. If you are doing things that are too far apart or don't have synergies then you won't see an impact. Ultimately, Jamestown's success is the result of strong civic leadership that created the plan and stuck with it over many years. This combination is rare. I don't think they completely know how unique they are.

Mayor Samuel Teresi, Five-Term Mayor of Jamestown (2000–2020)

We've been at it a long time. I'm a lifer here in Jamestown, born and raised here. I often say that longevity can be both a blessing and a curse. I am, with instant recall, able to put some history to where things came from and what has worked in the past—and not worked.

We hired a consulting firm to try and help us figure out what to do with the old railroad station. We got a Department of Transportation earmark, some CDBG money and foundation money and historic tax credits, and repurposed it for the National Comedy Center. Ultimately, the goal is to make it the Cooperstown of Comedy here in Jamestown.

People started getting engaged, and some of those people, nicely, were folks from the suburbs. They were saying, "What are the types of things we can do to the downtown that would make me and my neighbors venture there?" We had a lot of differences of opinion. But our goals were the same and we've elevated downtown and the Central Business District to its rightful place.

We've had a couple hundred units of new housing in the downtown area. Some of it, obviously, is affordable, subsidized, senior, and you don't want to fill your downtown up with that, you want market. And quite frankly, on the market side, we've kind of bought that new housing by stimulating the developers to take on hard projects with funding that the city has gotten from the government side or in many cases our foundations, led by our largest foundation, the Gebbie Foundation. It's about a $100 million foundation headquartered in Jamestown and their main mission is the redevelopment of the Central Business District of the City of Jamestown.

We don't have any Fortune 500 companies. M&T Bank is not going to come and underwrite our cultural arts program or our downtown development process. We don't have Warren Buffet and Geico or Jeremy Jacobs and Delaware North. But one of the things that we do have here is a lot of old philanthropic money. And how do we get that? It's caused some pain and resentment in certain sectors. The Gebbie Foundation transitioned out of the typical philanthropy like scholarships and grants to focus on downtown development. That created a bit of upheaval in the community and caused some readjustment but to tell you the truth, the foundation community is our equivalent of Jeremy Jacobs, and more important than just the money is that they have a stake. The foundations not only provide a forum for dialogue, but Gebbie can put the arm on some folks to get engaged in the game.

A Conversation

Bill Rice (BR), Jamestown Director of City Planning

Vince Joy (VJ), Jamestown Director of Economic Development

BR: The problem is we don't have the demand here. We had to rely on grants to do a lot of the projects but we were fortunate because we won a $10 million Downtown Revitalization Initiative from the state. That's a big part of affecting demand.

VJ: You have to work very hard to convince someone that that they should locate their 50-person software development team here in downtown Jamestown. Millennials need to feel that there are places for them to go to and that there are apartments in these old buildings with some adaptive reuse. You have to find people to actually come here and bring the jobs with them so that you have the restaurants and everything else around it. But we also know that this area is rich, and can be richer with tourism, with our lake, and the Chautauqua Institution, which is known world-wide. People from Pittsburgh and Cleveland and New York City and everywhere else come here for nine or ten weeks, spend an enormous amount of money to stay here and own houses here. But they want things to do and see that are different than just what's in that gated community. So we're trying to create an environment that's appealing to them too. **You've got to sell the sizzle. You have to build the picture. You have to be able to describe the vision**.

BR: We're fortunate that we can go from the downtown core in a short walk on a riverwalk to wetlands. We were just talking about, what can we do in the wetlands area to actually make it an attraction? By keeping it in its pristine condition and connecting it to higher education as a laboratory we have an opportunity. We have the Roger Tory Peterson Institute here. There's a lot of interest in that. So how do we take all these different components and link them together? We're even finding a way to get an excursion train here. The city put in half the cost and we've been going out to multiple counties in Western New York to help contribute because we see this as connecting all the way up to Niagara Falls, opening up that corridor to the Toronto market.

. . .

VJ: The National Comedy Center estimates it gets about 70,000 visitors per year and we're attracting the top comedy names. Whether it's Jerry Seinfeld or Jim Gaffigan or Lisa Lampanelli. They all have buy-in to the National Comedy Center. Stay away from calling it a museum. It's a center. It's basically a living, dynamic type of attraction that can constantly change.

With the Comedy Center comes additional private investment. We had a 1970s era Holiday Inn that's been mothballed for four months and I convinced a Buffalo-area developer to come in and purchase the building. Then there's an old Key Bank building, and they don't need this cavernous space. They ended up selling this gorgeous building to an out-of-town developer who has a summer home here at Bemus Point. I went for a Restore New York grant and we got $10 million out of the Downtown Revitalization Initiative. We're going to transform this beautiful old building that really had no use; a

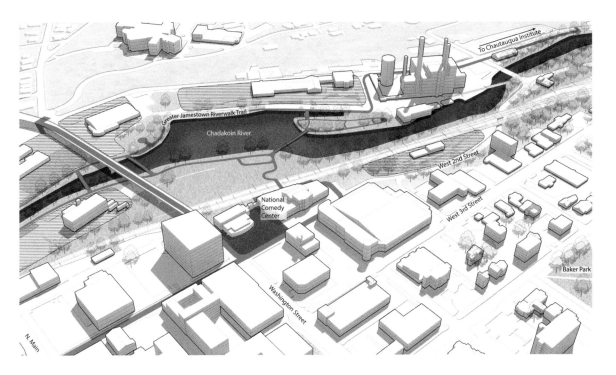

software company is going to inhabit it with fifty jobs. It will be commercial space, potentially a restaurant. There will be some market rate apartments. So it will be a truly adaptive reuse, mixed-use project that probably never would have happened were it not for the Comedy Center coming and the enthusiasm that's building around it and the potential that Jamestown is being transformed.

It was a $12 million project to do a complete rehab of this old railroad station—a Depression-era building that was basically abandoned and slated for demolition. Senator Clinton and Senator Schumer along with our Congressman at the time Brian Higgins were able to find federal and state transportation dollars. Governor Cuomo has been a big proponent of this. Just this year he slotted the last $5 million that was necessary. It's been a combination of local and regional private investment through foundations. New Market Tax Credits helped put it over the top.

We still need people moving into the city to basically keep our tax base up. Because every developer that wants to do some kind of adaptive reuse wants a tax PILOT (Payments in Lieu of Taxes). All this economic development activity for the city to maintain its operations, we're not going to reap much of the benefit because a lot of it is on PILOT. So we have to rely on our neighborhoods to keep that tax base up. If our neighborhoods keep declining it makes it more difficult for the city to do the things we had been doing to bring about this revitalization.

Basically, to go after any grants, we do everything we can in-house. We do all of the graphic design and all of the writing. We just don't have the funds to hire consultants to do any of this stuff.

BR: We're way outside of doing our regular work, but if we don't, we stagnate.

Journey Gunderson, Executive Director, National Comedy Center

In the late 1980s, Jamestown's art council, which ran the theater at the time, came to Lucille Ball and said, "Hey, we want to build a museum about you and your legacy." And she said, "Don't build a museum with just stuff about me in glass cases; make Jamestown a destination for the celebration of all comedy." I was hired as the Executive Director of the Lucille Ball Desi-Arnaz Museum and to bring to fruition Lucille Ball's bigger vision. The concept was a Comedy Hall of Fame, but that concept evolved based on conversations with the artists themselves who kinda felt like, you're Podunk Jamestown, New York and don't have a foothold in the industry to be deciding who's inducted and who's not. So our approach is much more that of an art museum. The Met doesn't say "We're inducting Van Gogh this year." You see the body of work and that's the emphasis.

We tried to design for the skimmers, the swimmers, and the divers. So, if you are a family taking a kid back to college and have an hour-and-a-half to kill in Jamestown, hopefully you come away knowing that George Carlin was significant even if you didn't know who he was before. But if you are a Carlin nerd, you could be here for hours *just* on this exhibit. Couldn't someone get that same amount of information from their smartphone in 15 seconds? Yes. So the onus was on us to make it more tactile and engaging. If someone is going to come all the way to Jamestown it has to be something they just can't find on the internet. When possible, we tried to tell a story with physical artifacts, many of which we got by reaching out to other institutions that don't have display space because they are not focused on comedy. The biggest consistent point of feedback we've had is, "Wow I really need all day here," which is key to the economic development piece. "Ok let's stay here and have dinner instead of getting back on the road" or "Let's stay and get a hotel room for the night and come back tomorrow."

People have compared this to Cooperstown, but you could find a lot of people who say "I don't like baseball." But you don't find nearly anyone who says "I don't like comedy." So I think that our chance of getting over 100,000 visitors a year is pretty good; we can expect to stabilize at 114,000. The resulting economic impact of that is $22 million on the region.

. . .

This building was called the BPU building, Board of Public Utilities, and a transformer was in this space. It was the trolley garage and repair shop. The lobby area and entrance we constructed. And then this hallway we built to connect the trolley garage to the train station, which is on the National Historic Registry so all the plans had to be meticulously reviewed and approved. And then at the same time renovation had gone into this building using New Market Tax Credits, because it had grown pretty dilapidated, and the Oishei Foundation was one of the founders. When they toured the community and walked through the train station and saw how beautiful it was they said "Wait a minute, why don't you make this the Comedy Center?" We didn't think it had the necessary square feet and so it took some time to figure out "Oh we could also acquire that building" which was owned by the city at the time and do a new construction to bring it all together to bring it to 37,000 square feet of exhibit space.

Before we raised the money to build the center we raised money through four or five different grants to build the park and seed and grade it and make it safe and make it so that you could cross the railroad tracks. There were a couple of trains a week, freight trains. And then there is a patio down here and we built an elevator shaft so that it's ADA [the Americans with Disabilities Act] accessible from the downtown. Without having to pay admission you can get down through this underpass or take the elevator down to the park. The next phase would be band shells, playgrounds, things to activate it. There's a river walk and now two new pedestrian bridges.

The space of the original train station we've made our gift shop because this space actually needs to remain public based on the funding that went into rehabilitating it. We needed every square inch for exhibit space so there's not a huge restaurant on site and that's part of the economic development as well—people get done and patronize the local businesses.

. . .

In 2010, it was Lucille Ball's one-hundredth birthday anniversary and the opportune moment to share the larger vision and say to the world "This is the track we're on." I was hired and I said let's make it a Comedy Festival. That festival, Joan Rivers headlined. It was so successful that the Chautauqua Region Community Foundation funded an economic impact study that quantified the economic impact from those four days. $3.7 million dollars poured into the County of Chautauqua and we had ticket buyers' zip codes from about 40 states. That helped us make the case that people would come year round for comedy. We used that study to get the attention of foundations to fund the Master Plan.

Even the backs and sides of buildings are adorned with homages to Lucille Ball and her beloved *I Love Lucy* show.

We knew from day one that this could easily become the butt of a joke if we did it badly. Comedians are very good at making fun of things. And then we're Jamestown, New York doing this? So I think from the start we had a very healthy awareness of who we were and weren't that helped us.

Dr. Lillian Ney, Board, Gebbie Foundation and Jamestown Renaissance Corp.; Former City Council President and Retired Physician

I'm from Jamestown originally. I was President of the Gebbie Foundation at the same time I was President of the City Council. And prior to that, I had been the Co-chair of the Strategic Planning and Partnerships Commission.

We started wanting to do an urban design plan. At the Gebbie Foundation we asked "What can we do to be a catalyst for something?" We decided to build an ice arena because we knew people in Jamestown wanted one because we had some retired Olympic winners here, Kirk Wyse and Lenel van den Berg. They were teaching at a city-owned ice arena up in Elm Park and they needed a lot more room. We felt it was really important to have it in downtown to show the revitalization process. It was almost $20 million dollars.

I was in charge of our annual retreat at the Gebbie so I invited Mark Kramer from Harvard. He said to us, "You're giving money everywhere, all the not-for-profits, every year. You've got to focus if you want to see difference." The then-president and the vice president, Chuck Hall and Rhoe Henderson, went out to every not-for-profit that we gave to and said, we had to decrease your allotment because we have to focus on economic development in the downtown. And that has been the strategic direction of the Gebbie Foundation since then.

We did an RFP to architects to refurbish some of the buildings: facade restoration, signage, awnings, whatever needed to be done. Nobody paid for it but us the first year. The second year the property owner had to give something. The following year the city was brought into it. And so we had a triple input: the city, the property owner, and then the Gebbie. Some of the buildings that were done on Main Street are really very nice.

We created a public–private partnership called the Jamestown Renaissance Corporation whose mission was to implement the urban design plan. After we did the first urban design plan, we said we have to have a neighborhood plan. Because the city would do a little bit here and there to make everybody happy. But doing a little bit in a lot of places doesn't inspire people because they don't even see it! So we got a neighborhood plan from CZB.

Nobody wanted to talk about the train station in Jamestown because as one of our board members said, it's a bottomless pit, it's going to be another 20–30 million dollars. How can you possibly take that on? We put a roof on just to preserve it, and decided we needed to get a transportation grant from the federal government. I got the idea we should hire a political consultant because how else could we get a federal transportation grant? Well damned if we didn't get the money in the end.

Facade renovation work in progress on one of the impressive historic buildings in Jamestown, New York.

One of the reasons we did the Arena was because the morale was very low. The city fathers couldn't understand that the downtown of yesteryear wasn't going to return. They said we'll get the same stores we had. They would not let that go. Well that was not going to happen and we knew that at the Gebbie. So when we did the Arena we had to take everything down that was there on that footprint. Then we relocated all the businesses that were affected to another, better place. We did upper floor development here at the Gebbie. We made the Wellman building, small, market rate apartments where a lot of people live who are consultants here in town or down at the hospital for six months. Young millennials have stayed there until they wanted to buy their house.

Patience and perseverance are absolutely key. I do a lot of encouragement of the young people here. It's very hard for them when they run against opposition and don't understand that it can't happen right away. You just have to wait for the right opportunity and make good relationships that are built on trust so that when some gap or potential is there everybody can work together on it. We've shown through the many years here that it can work.

Andrew Nixon, CEO, Chautauqua County Visitors Bureau

Lake Erie or Chautauqua Lake are what definitely sets us apart from a lot of other rural areas of New York. Lake Erie is generally clear, it's very nice. We do a thing where we subsidize a program that maybe sixty or seventy people go out on a half-day walleye charter. After three or four hours of fishing they go have a walleye lunch and then they have some meetings and talk about issues, and it's just an awareness builder between our county and Erie County. Always you bump into people that say "I just never knew how nice this was." One of our big issues is lake management. Chautauqua Lake and the smaller inland lakes have invasive weeds and when they die that contributes to algae development.

. . .

I think the state doesn't believe in the gateway centers right now. They haven't put any money into completing the last two that were previously in their plan. The state did decide to designate ten welcome centers without talking to anybody. Well guess what? We don't have one. There's the idea of putting information in front of people at strategic locations, but we're told "There's no money for that right now, we're not interested."

There's not much working together across municipalities or parts of the county. It's sort of a "my town" approach. Politically it's a challenge for people to reach outside of their municipalities, but it needs to happen. We need to rise the tide, otherwise you're just battling 1 or 2 percent growth, that's hard to *really* grow when that's happening. So you got a lot of municipalities, a lot of government, a lot of initiatives, and not enough working together.

. . .

Transportation is a challenge in respect to attracting people from areas like New York or Philadelphia, even Boston. People from those areas are probably going to fly to Buffalo most of the time and rent a car. But if they're like "I don't want to rent a car, can I get an Uber?" Well, that'll cost you $200 to get down here. "What about the train?" Yeah, there's no train. As far as the excursion, there are tracks behind the Comedy Center and there are some north–south tracks that are in disrepair. At the end of the day, I found in talking to railroads they can just say "Nope," and that's the end of it.

A Conversation

Emily Morris (EM), VP Communications and Marketing, Chautauqua Institution

Jordan Steves (JS), Director of Strategic Communications and Community Relations, Chautauqua Institution

EM: An organization like Chautauqua Institution is *very* important in this region, not only by virtue of the taxes paid by our property owners but also the impact of our nine-week summer season bringing some 100,000 visitors every year to the grounds. While we have a little more than 100 employees year-round, at the end of June through the end of August we balloon up to well over 1,000 employees.

There really isn't another place like Chautauqua, which makes it hard for people to understand who we are. It is part of the agenda of our current president and certainly presidents before him to show what a resource Chautauqua Institution is. That's both from a volunteer perspective—many of our staff serve on boards and leadership roles in organizations around the region, our colleague serves on the City Council in Jamestown—and through the arts and education outreach engagement programs that we are doing.

JS: Historically, there has been a bit of disconnect between Jamestown and Chautauqua. It has not always been easy to get back and forth. So we've actually worked with one of the local foundations to create a more regular link. The week we did on comedy recently, there was a regular shuttle between here and Jamestown. It would be great if it there was a circulator.

The County is working on the Barcelona Chautauqua Trail which would run from Barcelona (on Lake Erie) across Chautauqua Ridge and connect directly to us here. Jamestown just created a new river walk. The Comedy Center connects downtown to the waterfront for the first time ever so that people can actually come up from the lake, and they can have events out back of the Comedy Center. It's exciting.

. . .

EM: I think about this as a *region* of which Jamestown is a part. The Robert Jackson Center is an educational partner that we work with very closely. The Robert Tory Peterson Museum, the Audubon Nature Center, Panama Rocks. We have always worked with those organizations. The Lucy Desi Museum has

Scenes from the verdant Chautauqua Institution.

been an asset. The regional wineries are also a favorite. There's some tremendous assets in this region that have emerged in recent decades and have helped to make Jamestown a likely candidate for something like the National Comedy Center.

. . .

EM: Right now, the third most popular tourist attraction in the state of New York is in Corning the, Corning Museum.

JS: Corning, NY is amazing, but the Corning Corporation is responsible for all the revitalization you see on their main drag there. Jamestown doesn't have that one company.

EM: Right, but they have a much more sustainable system. I don't think it's good to be relying on one company.

. . .

EM: The fact that Chautauqua is a gated community in one of the most economically challenged regions of the state is a point of tension. The gates are only staffed after 6:00 at night and in the summer so that we can make sure you've paid the entry fee. They're not designed to keep you out.

I reject the idea that it's a tourist location, because people are taxpayers here, and while they may not spend their whole life here they do invest a lot of resources here and those resources benefit the region and always have. Our mission is education and enrichment. We used to offer degrees. We don't anymore, thank goodness, but we have an educational mission. Our programming includes the arts and interfaith investigation and recreation. So it's a mind, body, spirit sort of an enterprise.

Notes

1. 2021 population (U.S. Census Bureau).
2. In the 18th and early 19th centuries Jamestown was tied to these nearby population centers via waterways, and the arrival of the Atlantic and Great Western Railway (later named Erie Railroad) in 1860 greatly accelerated the distribution of both people and natural resources. The rail line linked New York City to Chicago and St. Louis through Jamestown, quadrupling its population between 1890 and 1930.
3. Although the terms are often interchanged, there is a difference between what a railroad calls a station versus a depot. A station is a physical location and can be nothing more than a mile-post with a name on it. A depot is the building designed for passengers, freight or both.
4. The mural artist Gary Peters received permission from CBS Inc. and Desilu, too LLC to replicate two scenes from the *I Love Lucy* series.
5. Refer to *The Lost Neighborhood Collection*, edited by Joan V. Cusimano Lindquist (2010) for detailed personal accounts of Brooklyn Square, a once-vital but aging business district that was demolished in the 1970s to stimulate investment.
6. Beginning in 2021, the state increased the amount to $200 million to support up to 20 downtowns across the state.
7. The recommendations from the workshop built on a 2015 AECOM Downtown Jamestown Economic Development Strategy report, the Downtown White Paper and other foundational studies. For the full list of resources and to the report, see: https://www.ny.gov/sites/default/files/atoms/files/WNYDRIAwards.pdf.

References

czb (2010). *Reinvesting in Itself: Report to the City of Jamestown, NY*. Jamestown: Jamestown Renaissance Corporation.

Cusimano Lindquist, J.V. (2010). *The Lost Neighborhood Collection*. Jamestown: Fenton History Center.

Halberstadt, H.A. (2002). *Train Depots and Round Houses*. St. Paul: MBI Publishing.

Hall, P. (2002). *Cities of Tomorrow: An Intellectual History of Urban Planning and Design in the Twentieth Century* (3rd edition). Oxford: Wiley-Blackwell.

Mumford, L. (1937). "What is a City?" In R.T. LeGates and F. Stout Frederic (2000). *The City Reader* (2nd edition). London: Routledge, pp. 92–96.

Rypkema, D.D. (2014). *The Economics of Historic Preservation: A Community Leader's Guide*. PlaceEconomics.

Swanstrom, T. and Judd, D. (2006). *City Politics: The Political Economy of Urban America* (6th edition). Taylor & Francis.

1.3
Bentonville, AR

Bentonville, Arkansas represents, in contradictory fashion, urbanism at the confluence of some of the strongest American traditions. These traditions include connections to our rural, agrarian roots; preferences for the scales and social life of small-town living; and importantly, participating in the technocratic flows of global corporate capital—if to extremes (Rowe, 2009). Bentonville barely qualifies as a town given that it has more than quadrupled its population between 1990 and 2021.[1] Over the course of researching this book, the population reached 56,734 people (2021), moving the town above our own qualifying metric of municipalities with 50,000 people or less. We elected to keep the case, nonetheless, because of the way Bentonville, anchored by the Walmart Corporation, highlights both the potentials and the polarizations of the public–private paradigm of American urbanism. Although many take the public–private city for granted, the model warrants closer examination in the Bentonville context where public goals are led and fulfilled by philanthropic largesse. The influence of major private corporations is particularly pronounced in Bentonville's Northwest Arkansas region (NWA), where global businesses concentrate in a politically conservative state with celebrated deregulation that supports them. The explosive growth of Bentonville is tied to this friendly political-regional-business context. Like the Intermountain West highlighted in this book, NWA is one of the fastest growing areas in the United States.

Beyond Bentonville, the region is home to a constellation of towns that are all growing together, each anchored by a major company or institution. Twenty-nine miles (46.7 km) from Bentonville is Fayetteville (with the large University of Arkansas campus). Springdale is just 19 miles (30.6 km) away (with the headquarters of Tyson Foods and its labor force) and Rogers, with Lowell nearby is 7.5 miles (12 km) away, known for the headquarters of the logistics corporation JB Hunt. Collectively, the towns in this constellation contribute in different ways to the ecosystem of corporate industry growth and labor, tourism and institutions.

Yet despite the intense growth of its towns and industries, NWA is within the sixth poorest state in the nation (Gilligan, 2023). It risks urban degradation, congestion and ecological deterioration amidst impacts of global climate heating that truck-based logistics, uninhibited consumer growth and physical expansion without public transit or environmental protections perpetuate. While the median income in NWA has grown along with jobs and the population, housing prices have also skyrocketed, creating tensions around affordability for those in the crucial service sectors upon which the industries rely (Golden, 2022). These stark realities open questions about the uneven nature of "growth and prosperity" that ultimately falls short of translating into broader *community* wealth.

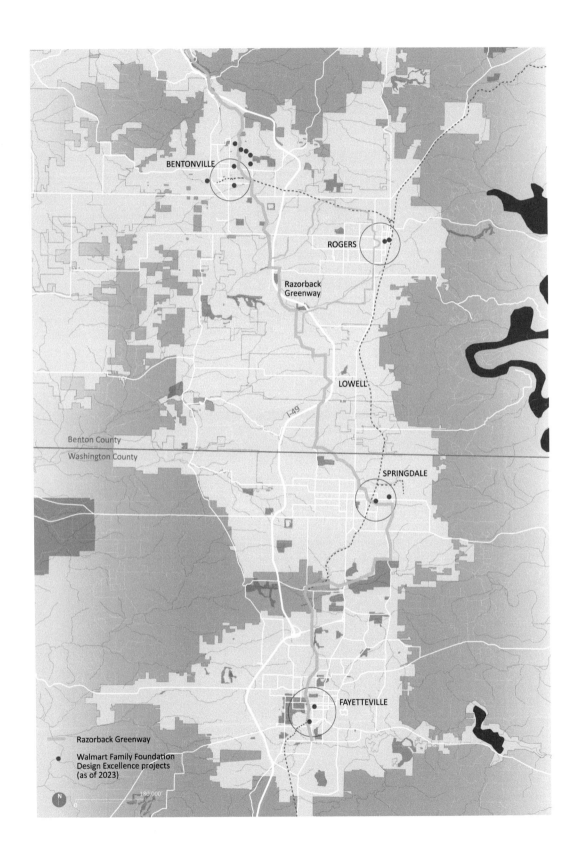

Walmart Inc.

In Bentonville, the historic presence and ongoing operations of the Walmart Corporation's global headquarters has defined the town. One of the original Walmart five-and-dime stores founded by Sam Walton—now a museum—fronts the central public square in Bentonville, a quaint downtown with a walkable urban grid lined with two-to-three story buildings and a prominent court house. Over the years, the five-and-dime model was revolutionized, developing into the big box typology of high-volume superstores with "just-in-time" barcode-enabled logistics witnessed everywhere. In small-town America, Walmart is often demonized as the economic bully that syphons off the viability of locally owned Mom & Pop shops. At the same time, many municipalities vie for the big box retailer to reap taxes from the high volume of sales they produce, absorbing customers from a regional catchment—even as those typologies inhibit walkability and erode the character of streets around it. **If Walmart has a reputation for killing the small town and main street retail, then in Bentonville, through its headquarter operations and extended foundations, the company *remakes* it.**

The Walmart Corporation is immense. With over 10,500 stores in 19 countries, the retailer is over five times larger than its next closest competitor, Target. It has been the world's largest company by revenue for the last decade. The company's influence affects the town not only with its physical footprint and countless vendors and employees, but also via the interests and capacities of the Walton family heirs who have committed to the region. By virtue of locating their corporate headquarters in town, and the concentrated philanthropic activities of two generations of Sam Walton's family, NWA has become a nexus of population growth, major cultural institutions, new urban public space and recreational projects that reach well into the landscape.

Interestingly, local opportunities for *high design* emerge from these philanthropic activities, made possible where civic capacity had otherwise not been able to manifest. Like many early 19th century company towns that preceded it, Bentonville offers cautionary lessons about the evolving role of corporate industry

From small-town storefront (now the Walmart museum, under renovation in 2023) to global logistics largesse: Bentonville is home to retail giant Walmart, putting Northwest Arkansas on the world map.

in making place—a role that brings funding along with market logics, branding, labor recruitment—some say exploitation—and global financial wherewithal.

Corporate Footprints

As the Walmart Corporation grew over the decades, its Bentonville headquarters spread out across buildings and blocks in town. In 2023, construction was underway on a massive new consolidated, 350-acre headquarter campus in Bentonville. With a master plan designed by Boston-based Sasaki Associates and buildings designed by Gensler, upon completion it will employ 14,000 people. Literally and figuratively, the campus is larger than the footprint of the historic Bentonville town center. The campus sits on the edge of that town center, between residential neighborhoods, a commercial strip and the interstate. The sustainability goals are ambitious: campus energy plants will be all-electric, and many of the buildings are constructed in locally sourced mass-timber (Repovich, 2023). The construction so far has diverted 95 percent of its waste from landfills, as it captures storm water and will encourage employees to bike and use electric cars (Walmart Corporation, 2023). The project has re-centered Arkansas' bountiful wooded "fiber belt," a local resource for construction lumber in lieu of more carbon-intensive concrete or steel imported from farther away.

Part of Walmart's purchasing power has to do with its control over all aspects of the (still carbon intensive) supply chain. Vendors of a certain scale must establish a local office. And the company controls its own vast transit logistics. This power of influence extends beyond the new corporate campus and company operations and into cost-saving controls and limitations on architecture, employee hours and benefits worldwide (Bonacich and Hardie, 2006). The company's foundations and heirs meanwhile imprint another kind of urban rebuilding that brings design to the NWA region to attract and retain Walmart employees and vendors.

Many major corporations are recognizing their role in addressing climate change. For some, this includes implementing stringent new sustainability goals and finding ways to outwardly model healthy environmental values. For (far fewer) others it might involve a whole-scale reassessment of the company's own root systems perpetuating and aggregating environmental harm. How do we ultimately measure a benefactor-company's commitment to the health of a town and to its own campus when that corporation's business model is an underlying reason things like local economies and the broader environment are eroded in the first place?

Foundations of the City

The Walton Family Foundation (WFF) was established by heirs of Sam Walton, the founder of the Walmart Company. Sam opened the first "Wal-Mart" distribution center and General Office in Bentonville in 1971 after testing his retail model in neighboring Rogers, in 1962.[2] The Foundation today continues as a national entity, but with attention focused on NWA. Broadly, it advances a mission that "tackles tough social and environmental problems with urgency and a long-term approach to create access to opportunity for people and communities" (Walton Family Foundation, 2022a). The foundation's local investments in NWA, or what it calls the "Home Region," make up one-third of their wide-ranging pillars (Walton Family Foundation, 2022b).

In an environment so defined by the efficiency logics of "big capital," it is hard to imagine that any town, city or civic sector business model would have room for the kind of premium that "good design" costs. A local designer told us, "You have these kind of frugal billionaires that still—even at the second or third generation—pinch pennies and demand free services from designers."[3] But also, Arkansas is a politically conservative, heavily deregulated state. Another local planner emphasized, "Arkansas doesn't have the political infrastructure to create a kind of 21st century place, so any advances that are made are usually made through the market."[4] This means that design quality is usually one of the first elements to be eliminated, as added costs rarely fit the American austerity (corporate efficiency) paradigm. Here again Bentonville models that rule, and the exception. Good design—for public spaces no less—is advanced and subsidized by the non-profit WFF's Northwest Arkansas Design Excellence Program.

The Northwest Arkansas Design Excellence Program is one of the WFF's signature home region initiatives and its impacts are visible all over Bentonville and surrounding towns. The program has become a catalyst for elevating design and for funding various projects in the region by providing grants that pay for top or emerging designers (largely from outside Arkansas) to play a role in rebuilding. The focus is on major buildings and open space projects that support its mission. Thus the public sector is beholden to the discretion of private capital (that it relies on), whether those whims match public needs or not. Nonetheless, the notion that Bentonville has emerged as a creative arts and culture hub with world-renowned venues, an exclusive 21c Museum Hotel, and other investments to draw in hospitality taxes is noteworthy and visible throughout the landscape (Souza, 2016).

The Design Excellence Program's built work has expanded impact in the region. The work serves as demonstration projects for municipalities who can

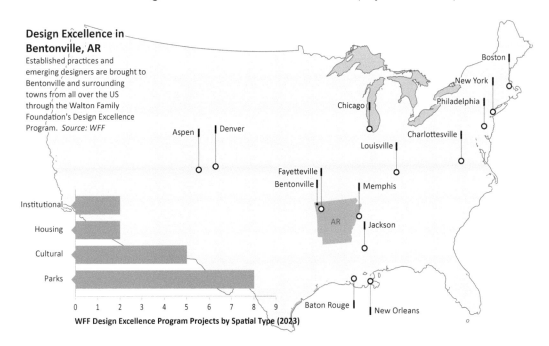

Design Excellence in Bentonville, AR
Established practices and emerging designers are brought to Bentonville and surrounding towns from all over the US through the Walton Family Foundation's Design Excellence Program. *Source: WFF*

WFF Design Excellence Program Projects by Spatial Type (2023)

then gain broader public support for, or recruit additional funding for, related efforts. More directly, the program has made new, literal city centers: It enabled the development of Springdale's administrative and criminal justice center. And a new town hall for Bentonville is in the works (Northwest Arkansas Design Excellence Program, 2024). The Design Excellence Program elevates capacity—of public-sector governments, and of design as a core project principle.

The scale of this philanthropy, and the urban interventions it props up, has arguably not been experienced since the industrial age in the United States when magnates like Andrew Carnegie funded beautifully built libraries and other institutions. Places like Columbus, Indiana (pop. 50,391)[5] come to mind where, since the 1950s, the Cummings Foundation has invited leading architects to design buildings and landscapes in the small town. Still, as most Carnegie libraries endure as magnificent monuments of public architecture, the steel industry he created was critiqued for the social problems it reproduced, in labor exploitation and environmental harms, among other issues. It's hard not to consider Walmart's family foundation legacy in contrast to the architectural and environmental legacy of the Walmart Company that shifted city centers and retail space into car-oriented, disposable big box commodities. These public–private relationships embody U.S. policies of deregulation in exchange for philanthropic generosity. With the Design Excellence Program, the heirs of the foundation now prop up ideas of architecture and the public realm as a long-term investment with value that grows over time in sync with the community.

Art Culture as Catalyst

Elevating design and the arts has been a long-term effort. One of the most well-known projects of the Walton family heirs arguably put Bentonville on the cultural map. The Crystal Bridges Museum of American Art (completed in 2011) was endowed by the WFF and founded by Sam Walton's daughter, Alice. Nestled in the Ozark forest just a short walk from the main square of Bentonville,

The Crystal Bridges Museum of American Art in wooded context in Bentonville, Northwest Arkansas.

the museum attracts more than 300,000 visitors annually from around the world to see work by the biggest names in American art—*for free*. Designed by celebrated architect Moshe Safdie, the collection of low-slung "bridge-like" buildings are organized around a stream and pond. The complex has become such a draw that the museum is undergoing an expansion that will increase its size by 50 percent. Scheduled to be completed in 2024, the addition will add classrooms, outdoor education spaces and larger exhibition halls.

Like MASS MoCA in North Adams highlighted in this book, with its context that mixes a gritty post-industrial mill town with the rolling Berkshire Mountains, Crystal Bridges functions as a cultural catalyst whose allure comes, in part, by the unlikely setting in which it is found. Crystal Bridges leverages its relationships to the 120 acres of grounds around it. Five miles of trails connect from the museum to additional private institutions sponsored by the WFF: These include the Scott Family Amazeum and a holistic health center; the Helen Walton Children's Enrichment Center; and further down the road, the Thaden School campus, an independent secondary school (currently expanding to include primary school students). Another spin-off institution includes the Momentary, an arts-incubator and concert venue a few blocks away.

Main Streets

Aside from the buildings described above, the foundation supports small nonprofit/retail type enterprises to get a foothold in, of all places—Main Street. The foundations of the other regional corporations do the same, as door signs on the facades of various businesses call out these subtle threads of subsidy. Naturally, it is in the corporations' interests to draw people to the towns and many small retailers struggle to compete not only with big box retailers and spiking rents but also with the increased market share of e-commerce.

Restoring Main Street today requires that municipal governments rethink their relationship to retail as part of broader rebuilding agendas. This has led to creative new strategies visible in NWA that return to ideas of the town center as an important place for public gathering, of which retail is just a

The remade downtown public spaces of Rogers (left) and Springdale, Arkansas (right), both enabled by the WFF Northwest Arkansas Design Excellence Program.

1.3 Bentonville, AR

part. The town centers of Rogers and Springdale have both been boosted by WFF support to complete major public space/parks to activate their main street areas and intersect with the regional inter-town Razorback Greenway trail system. The quality of the streetscape infrastructure is notable.

Happy Trails

Other members of the Walton family have imprinted their legacy on Bentonville through entirely other means, building a network of mountain biking trails immediately surrounding the town. These trails take advantage of the NWA region that is nestled in the heavily forested Ozark highlands. As the cultural institutions, parks and public spaces attract future vendors and employees (and tourists) to the region, so too do the trail systems. Bentonville has become the "mountain biking capital of the world" for enthusiasts. This claim to fame brings mountain bikers from all over the United States of course, but the recreational amenity even extends into new resident and worker recruitment programs that celebrate healthy living and the outdoor lifestyle. As an example, in 2020, the non-profit Northwest Arkansas Council promised every new resident a free bike and $10,000 to relocate to NWA (Annis, 2020). Bentonville's small downtown connects by trail directly to its leading cultural institution, the Crystal Bridges Museum, but also to the Slaughter Pen network of mountain bike trails and others to the north. **If downtown Bentonville is becoming a global arts destination, it is also a national gateway to the surrounding larger landscape and recreational network.**

Bentonville's bike trail network radiates outward—literally a block from the main square downtown. The dirt trails are run by an organization called OZ Trails. The paths are engineered with sculpted tricks, jumps and turns, and labeled like ski slopes with fun monikers by level of difficulty. They intersect the Razorback Greenway, a paved, more generally accessible path that was started in Fayetteville over twenty years ago by the public Northwest Arkansas Regional Planning Commission (NWARPC). In 2015 it was completed, reaching Bentonville 37.5 miles (60.4 km) later, with help from Walmart (NWARPC, n.d.). More than just a recreational asset, the economic benefits to the region of the cycling culture and infrastructure are significant, estimated at $159 million (Sparkman, 2023).

Not surprisingly, when one is in Bentonville, it is very hard to tell which systems are publicly versus privately funded or owned. They are all free and open to public use. Is this clouded ownership a benefit, or symptom, of the erosion of public capacity in American cities? The protracted and painful shrinking of municipal budgets has required private-sector involvement to compensate for, or at least augment, services that towns traditionally provided. But apart from the greenway (which is significant) the regional governments may not have indulged spending on such a massive recreational system of mountain bike trails for what could be considered a relatively narrow user group. But hardcore mountain bikers, recreational cyclists and residents alike love these trails, and the town thrives on their tourist dollars as well as the commuter connectivity they provide. In fact, the entire town has become bike-friendly. With Walmart and Walton family heirs as sustained patrons over generations now, Bentonville celebrates amenities most small towns can't afford.

Biking culture in Bentonville spawns creative new urban typologies: A new residential development inverts its entries away from the road and toward its mountain bike trail.

A new commercial building called the Ledger boasts complete bikability, via a three-quarter-mile ramp that encircles the entire exterior (and more on the interior).
Credit: © Tim Hursley.

Much like the Beltline in Atlanta and Railroad Park in Birmingham, Alabama, the trails close to town spur new residential development (Gamble and Heyda, 2016). The economics of this model have been critiqued as creating gentrification and pushing low-income residents out. But architecturally it offers new ways to rethink typology, as buildings now accommodate *two* fronts, one to the street, and the other facing the greenway or mountain biking trails. Everywhere in Bentonville business and design cultures accommodate bikes as an alternate form of transit, and this is an exciting, unexpected feature of the area. A mountain bike trail runs alongside new downtown trailside development. In another part of town, the Ledger City

Bike culture in town defines a new kind of "on-street parking" exclusively for bikes outside a local bar.

Center building boasts complete bikability. A gradual ramp encircles the 230,000 square foot office complex completed in 2022 that allows riders to ascend six floors to a restaurant and bar terrace on top—without ever getting off their bikes.

Apart from mountain bikers, recreational bikers and commuters of all ages can be seen throughout the town. For a region that has stubbornly refused to prepare for or coordinate mass transit, the bikes are a refreshing, sustainable push away from cars. The presence of so many people moving freely around the town on bikes permeates a spirit of relaxation that makes the land-locked Ozarks hamlet feel like a casual beach town in the summertime.

Routes to Transit

In many American towns provisions for public transit—light rail, in particular—are arduous propositions for local leaders without access to sufficient revenues for construction or operations. Small geographic footprints and low-density settlement patterns make it hard to obtain or sustain the optimal ridership needed to cover costs of building and operating fixed transit systems; few towns have them. In many towns even a bus system is lacking (although in tiny Greensburg, Kansas you can use a free bike to get around). But as the Bentonville region grows within the constellation of adjacent municipalities, so do the main arterial roads and highways. Mass transit makes sense here given the population growth and the urgent imperatives to transition off fossil-fuel dependent car-based mobility, a leading cause of global heating. Planning studies have shown how and why transit should be implemented in the region for both ecological and social-economic benefits (University of Arkansas Community Design Center and Washington University in St. Louis Sam Fox School of Design & Visual Arts, 2007). Many workers in the leading industries live in adjacent towns and commute. Unfortunately, achieving efficient transit in less dense U.S. settings, let alone in NWA, seems an impossibility as long as the governments lack the political will and capital to invest in it and especially as the private corporations supporting the towns and their employment base appear fine with the status quo.

Some may point to the perception of transit and trains as a threat to Walmart's road and truck-based logistics business model. Others note the conservative political leanings of state actors who see rail as a threat to anti-public "freedoms" and individualism, ideologies that highways—personal cars—are believed to foster. If a critical mass of elite "stakeholders" does not support progressive systemic urban change, it is hard for local leaders to push it forward, especially in a small-town setting. Moreover, when the "stakeholders" are unevenly made up of the area's, if not the world's, largest corporations who oppose the idea, it is harder still. Sadly, even in the most urbanized areas across America there is ambivalence or outright resistance to spending on public transportation. In this way, Bentonville represents a microcosm of the kinds of American planning tensions that exist everywhere: tensions that pit the good of the collective and health of future generations against ideas of individual liberties that defend the profitable systems that stymie change. As if to cement the reliance on personal cars well into the future, an astonishing number of hulking new, massive parking structures have also been constructed in Bentonville alongside many of the new projects described above.

A Microcosm of American Urbanism: Global Towns, Private Actors

Bentonville represents an American town more global than parochial. If cases like San Ysidro in this book represent localized actions on the ground by a non-profit organization to help a disenfranchised community in the face of massive globalization infrastructures; Bentonville also represents local design capacity and action led by non-profits—but also industry behemoths. The consequences in Bentonville are that investment decisions build local capacity by reproducing corporate wealth, so they inevitably don't address inequality. Bentonville's changes are in the context of ongoing major global retail upheavals that are diminishing walkability and thwarting small-scale enterprise. In Bentonville, design and city-making contradictorily stems from the same root (corporate) systems causing upheaval in the first place. Urbanism by philanthropy reinforces a loss of democratic public planning ideals in American cities, but celebrates high quality, if uneven, access to amenities and programs granted by the whim and generosity of private or non-profit actors.

Perspectives

A Conversation

Karen Minkel (KM), Former Home Region Program Director, Walton Family Foundation; Philanthropy Director at Lever for Change

Kathryn Heller (KH), Senior Communications Officer, Home Region Program, Walton Family Foundation

KH: When we're talking locally and regionally, we are an urban area. But when we talk to a national audience, to national media, we're very much a rural community. Here in Arkansas, because most of the state is truly rural, it's hard to say that without putting Bentonville on the same playing field as, say, Gravette, Arkansas. That's not fair to the people of Gravette because we don't have the same resources. But when you're talking to a reporter from the *Washington Post*, especially if they visit, and they see cows off the side of the highway on the way in …

KM: When we first set up the Design Excellence program, it was with this intent of: How can you elevate the design of public buildings and spaces in the region? What I don't think we realized was how the design program would really serve as a tool for organizations and especially municipalities to garner public support for a project. In the third year, we wanted to solicit public space projects, because up until then, we'd really been focused on buildings. We did an open call. Any organization could submit a public space project. But the projects that were selected were all downtowns and three of the four were submitted by a municipality. The visions were all to create gathering spaces downtown to promote interaction among different groups of people. After we supported their projects it became evident that it was incredibly beneficial for the cities to have conceptual designs when they went forward for bond initiatives. I believe three of the four went out for bond initiatives *after* these designs were at some level of completion, and used that as a tool to generate public support for the project. That was something we hadn't quite anticipated.

In the recent evaluation, we were looking to see what had been invested in terms of public support in these projects. Over $100,000,000 in additional public and private funds for the construction and development of Design Excellence projects had been implemented. That's pretty significant.

TheatreSquared (in Fayetteville, AR) did something similar. We provided some capital support for the actual construction of the building, but they were able to raise a third from public donors and a third from other

WFF Northwest Arkansas Design Excellence Projects: (left) Railyard Park in Rogers, AR by Ross Barney Architects (Chicago, IL) and Nelson Boyd Woltz Landscape Architects (Charlottsville, VA/Houston, TX), (right) Bentonville's Helen R Walton Children's Enrichment Center by Lewis Tsurumaki Lewis (New York, NY).

funders. And they were also able to share concepts. I'm not sure that these projects would've come to fruition otherwise.

. . .

KM: I would add that we really needed some of the recognizable names in the pool to sell the program. I think in our last round the program was at the point where it was bestowing some recognition on an up-and-coming firm and enabling them to get a significant commission that would then promote them in the broader field.

. . .

What I've seen across the region is the role of philanthropy in catalyzing a lot of the same investments and thinking across all the communities. The focus on downtowns was inclusive of public spaces and cultural facilities. There's also been real attention to how people move around and how you can create more walkable downtowns. Bentonville had also been working on a trail system and with a push from the Foundation, that trail system now spans the entire corridor. So it became a 36 mile (58 km) spine called the Razorback Greenway and then a network of trails was created from that.

. . .

And there have been field trips. We worked to take a group of elected leaders and officials to Copenhagen to experience what it's like to have a city with a completely different infrastructure and with a different set of assumptions about how people are going to move around. And the mayors in these towns have all become champions for walkability and bikes, so they're redesigning streets. The Foundation has been catalytic in providing funding that has enabled elected officials, staff and community members to learn—Copenhagen was probably the most dramatic example. But we have also supported a series of workshops here, where people who are doing things like

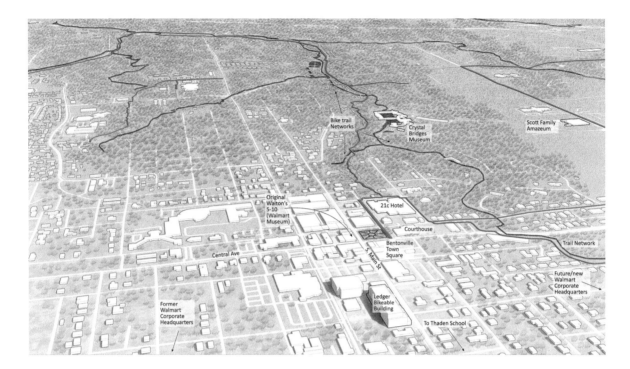

transportation in Kansas City, Oklahoma City or Indianapolis have come in and talked about how they've been successful in a push for public dollars in a primarily red city or state or county. And then we've also supported the catalytic infrastructure that has given elected officials some political cover to try something different, to design a street or test something in a new way.

. . .

There was an article recently about funders doing this across the country – if you look at Tulsa and the George Keizer Foundation, or Kansas City and the Kauffman Foundation, and even New York with all the projects in the city being catalyzed by a significant funder. I see that same thing happening here, but because of the region's size and the scale of the Walton Family Foundation, you're seeing it happen week-to-week in a really visible way that isn't captured in the same way in a large urban area.

. . .

The region has built some strong collaborative muscle; they've come together for various projects, so I think there's just a spirit of cooperation. When the Razorback Greenway was built, it leveraged TIGER II funds[6] and I remember Lynn Richards (President and CEO) at the Congress for the New Urbanism said something like "a lot of communities have trails – but collaboration across six municipalities to get this significant piece of infrastructure in place in a relatively short amount of time… that's the story." I also think **there's a recognition that we're growing so quickly that you really need to cooperate in order to manage that**. The foundation is also helpful in that regard because it provides resources, so there isn't a feeling of scarcity among the different cities. There hasn't been a feeling of anything beyond friendly competition. Which could always change. **It's a big question what Northwest Arkansas looks like 20 years from now, whether it becomes imperative that it is all one municipality or municipal entity.**

Stephen Luoni, Director and Principal Designer, University of Arkansas Community Design Center (UACDC)

We [the UACDC] proposed a light rail project fifteen years ago. Heavily researched, we consulted national authorities on it. We went out to Portland, met with all the people that do street scores and transit; Berkeley, CA with all the transportation engineers, and they said this makes sense in Northwest Arkansas. We did a book on it. And the political establishment went to try and get the report stopped.

. . .

At the level of the region, you have to triangulate solutions to housing, transportation, and land use. And our light rail study was really about that, because this is one of the fastest growing regions in the country and it's developing at a very low density. Ironically, Bentonville, I think because of the wealth, and because the need to aggregate by a creative class that can afford it, is one of the few cities that is actually densifying as it grows. But in the region it's a very different story. Because we don't have policy, we'll never be able to address sustainability, we can't get a handle on land use. The average Northwest Arkansas household spends 29 percent of its household income on transportation. The national average is 19 percent. In cities serviced by rail, the average household income on transportation is 12 percent. We make the argument that if you just get the transportation and land use correct, you could build prosperity. Transportation is key to downtown revitalization, prosperity, but also to sustainability and keeping land use checked. But the only problems we can solve for in the region right now are building airports, infrastructure, logistics, but not complex public-sector problems. I think that's the big hurdle for the next generation of leadership in Northwest Arkansas.

. . .

This has been an amazing region of the creative class, but that was never evident until about fifteen years ago because people quietly did their work, and then they flew out. It was like a military operation. But since the sixties there's been a creative class in Bentonville. Saatchi & Saatchi, the great PR advertising firm out of London—their largest North American office is in Springdale. They have like a 150 people there or they used to. And with the Walton Family Foundation, they have around 100 people that work there now. And there's Crystal Bridges. Bentonville's become saturated with cultural and creative assets for a town of its size. It's not the scale of Columbus Indiana yet, but it's certainly headed that way. Per square mile, the creative talent of Bentonville is just off the charts.

Marlon Blackwell FAIA, Principal, Marlon Blackwell Architects (MBA) and E. Fay Jones Distinguished Professor at the Fay Jones School of Architecture and Design at the University of Arkansas

We call it *Ruburbian*: Where the rule-set of suburbs begins to mix with rural and agricultural. We're looking at town squares and how public space is

The "bike barn" in Bentonville, Arkansas by Marlon Blackwell Architects (2021) exemplifies their exploration of a "ruburbuian" architectural vernacular that is at once contemporary but based on the striking forms of rural/agricultural buildings, and yet low and set back from the road in suburban fashion.

leveraged in small towns, but also factoring in how farm buildings are grouped together to create space and place.

We're looking at how chicken sheds are arranged in the landscape, because this is a big part of the chicken industry. They're long and narrow and linear and their long axis is east, west. That minimizes the sun exposure for the AM and PM, and they control it on the south, and catch the breezes, so the chickens don't cook before their time, so to speak. There's an environmental lesson there that we factored into how we organized the Thaden School campus. It's part urban in that we use the town square as the unit of measure and in organizing a building for the environmental response, for day-lighting and heating and cooling, breezes and such.

. . .

In many ways, it started with Alice Walton and her decision to build the first great American art museum since the Whitney in, of all places, Bentonville, Arkansas. I've talked to people in New York and San Francisco who were visibly angry about this, saying, "whoop-de-doo she thought she could get 250,000 people a year to come, we could get millions." And they missed the whole point, which is to get people who wouldn't normally go to the Whitney or to that new San Francisco Museum; by making it free and making in the middle of the country. It's a game changer, the development of this notion that art, nature, architecture should be available to everyone, that it is not exclusive.

Alice Walton brought a Frank Lloyd Wright house that was going to be sold off, maybe even destroyed, to the campus. [The Bachman-Wilson house was moved from New Jersey to the grounds of the Crystal Bridges museum.] She brought Buckminster Fuller to the campus, large scale. You started to see a change in clients. We could now say, "there's a whole conversation going on, and your project could be part of that." I think there was a repressed desire to have that conversation.

The Thaden School in Bentonville, Arkansas by Marlon Blackwell Architects; a project supported by the Walton Family Foundation Home Region Design Excellence Program.

There is a historical legacy of design excellence in Northwest Arkansas: E. Fay Jones; Edward Durell Stone, who is from Fayetteville. Warren Segraves, who was a "Miesian," did 120 buildings here. And the Ozark moderns are about eight or nine architects that did work here. I joke around that it's much harder than what Sambo Mockbee had to do in Canton, Mississippi. We're just standing on the shoulders of a lot of these folks here and taking it to another level. What we're doing, I believe, is making it more publicly accessible, because we decided [at MBA] to shift away from residential work, private residences, and focus on public work, institutions, and education. But having the opportunity of the Design Excellence Program to test ideas means we can then leverage that for work we're doing elsewhere in Dallas, in Detroit, in Memphis.

[The Design Excellence program] is allowing for smaller startup firms to emerge. They're getting recognition. The Architectural League of New York selected Fayetteville's *Modus* as Emerging Voices winner in 2018. The program has created notice of this area. Dave Wells out of Springfield is starting to do work down here. And El Dorado out of Kansas City is doing work now in Northwest Arkansas. It is a coalescing of institutions and businesses, like Walmart, like Tyson, like the University of Arkansas, who sort of changed their ways. And so it starts to snowball.

Peter MacKeith, Dean, Fay Jones School of Architecture and Design, University of Arkansas; Chair, Advisory Committee for the WFF Northwest Arkansas Design Excellence Program

Northwest Arkansas is probably heading towards being categorized as a metropolitan region of small towns, more so than any single urban center. I've always understood it as this archipelago of five or six small towns that each began and want to continue to be understood as small towns, and in many ways are struggling to remain small towns, with a town square, or center, or park, rather than simply an endless, seamless sprawl.

I know that you're looking at single-name towns or cities, and up until probably ten years ago, Bentonville and Fayetteville, and for that matter Springdale, Rogers, Siloam Springs, would've all probably raised their hand and said, "Please include me." But increasingly the perspective of the Northwest Arkansas Council—and behind that, the Walton Foundation, and [the] Tyson Foundation, and the other philanthropic/economic actors in the region—will all want to understand *Northwest Arkansas* as the place name.

We all know that Arkansas, like Louisiana, like Mississippi, like Alabama, is always occupying the bottom four or five placement of any national measurement of indicators of health, income, quality of life. [But] one thing that I had not really recognized was that Arkansas is one of the most forested states in the United States. Its 56–57 percent covered in forest and it's part of a belt of the southeastern states that is increasingly understood as the current and future "fiber basket" of the United States.

People in the timber and wood products industry simply think of trees as fiber, an organic material that can be transformed in different ways. So this is an immense natural resource for the state. All these southeastern forests are very much part of those economies and part of the environment. Arkansas' forests, similar to the other southeastern states, have been well-managed over time. They are almost super-healthy forests. They're overgrowing themselves to the point of danger, such that currently Arkansas is in a surplus biomass condition of about fifteen million tons per year. For every one tree that Arkansas cultivates and harvests and sends to the saw mill, there's not quite two trees that are growing against it. We just can't keep up with the productivity of the forest. And there's real environmental dangers there—wildfires, and insect infestation, and problems with the water table if wood is just left to rot.

It would be of environmental and also economic value if further uses for that surplus bio-mass could be identified. It was very clear to me that we [the School of Architecture] could demonstrate the increasing use of timber and wood products in sustainable design and construction. But the economic benefit came in 2019, when StructurLam, a Canadian manufacturer of laminated timber products, determined to set up its own cross-laminated timber factory in Conway, Arkansas, which is just north of Little Rock, to manufacture glue-laminated products and cross-laminated panel products out of Arkansas' southern yellow pine. Now, we could have a hundred of those plants and barely touch fifteen million tons of surplus bio-mass, but it's a start. And that plant is new economic development, so new jobs, and new tax-monies, and so on. It's somewhere around 500 jobs total, and it's a $90 million to $100 million investment—but half of that has been underwritten by Walmart, who determined around the same time, through some advocacy I was making, that if they were going to build their new headquarters (which they're now doing in Bentonville) and be true to their projected image as a sustainable and even regenerative company then they might want to think about building their new headquarters out of mass timber. They have decided to build their three million square foot headquarters out of mass timber and to source it out of Arkansas. I introduced them to StructurLam, and a marriage of real value was created.

Mass timber buildings under construction in 2023 at the Walmart global corporate headquarter campus in Bentonville, Arkansas.

This has nothing to do with small towns, but it has to do with rural communities in the Southern portion of the state, in particular the Southeast and Southwest, which are the forest-based communities and the sawmill-centered communities. They have been languishing for many years, as the paper and pulp industry has increasingly shifted to China or elsewhere.

. . .

When you think of Walmart, it's both local and global, in the sense that it is clearly devoted to where it is in Northwest Arkansas, and it is clearly reliant upon its markets in Mexico, Brazil, India, and China. They have whole backrooms of logistics people who are either first-generation or direct immigrants in order to communicate with their partners in China, Mexico, Brazil, and so forth. The more you move in and around Bentonville, the more you begin to see the diversity of the population. There's now a very clear agenda that's been brought forward through the Walton Foundation, and through the Crystal Bridges Museum, to emphasize that America is, in fact, a much more diverse nation than has been recognized. And that we in Arkansas need to recognize that through and through. A lot of the cultural programming that is being funded through the foundation, and through Crystal Bridges, has increasingly tended towards that greater and greater cultural diversity.

. . .

There are guiding principles for the Design Excellence program. Some of them do have to do with building economic capacity. But others have to do with working within a regional sensibility of architectural excellence. Design is not imposed upon communities, it is built from communities, and, so, it's as much about a process of community building around a project as it is about the building that is purposed towards the community. That has become a very important dimension for me to impress upon people who want to submit their credentials. You can't just show up into small-town America and say you've got the answers, right? That would be one generalization that I might propose, that: it's not that small towns are provincial; It's that they're very proud of what they've been able to do, and who they are.

. . .

The public–private interaction [in NWA] is also interesting because of the strong presence of truly well-endowed major corporate actors. And there's an increasing number. Canoo, this electric vehicle manufacturer, is showing up now in Northwest Arkansas, as well as across the border in Oklahoma. People are starting to arrive into this area because the population is growing and because there are other corporate partners to lure them here.

If Amazon is situating itself in cities, and saying, "You, young, new ambitious worker, should come to the big cities," to Seattle, or Washington, D.C., or New York, whatever. Walmart says, "Nuh-uh, we're going to stay right here where we are. Our proposition to you is that you want to live in the fresh air, close to biking trails and a great place to raise your family."

If Walmart wants to build on a site in the middle of the city, that's what it's going to do. And the City Council goes, "This will be great for the city." And it may well be. Those types of more typical municipal decisions are affected by what's good for Walmart at the same time. If you move down the corridor to Fayetteville, it starts to shift, and then the larger actor is the University, which is the largest employer there. Fayetteville has a history of being much more of a country town and far less economically centered. Orchard farmers more than five-and-ten store operators. All these other places, Springdale and Rogers—Rogers is J.B. Hunt, and Springdale is Tyson. There surely are ways of measuring the presence of those corporate actors within municipal decision-making, and property development, and who makes the planning decisions. We're at the early stages of discussing a curriculum for a Master's degree in Planning. But, interestingly, the call to us as a school, now, to develop a graduate degree in planning, is coming from both the public and the private sectors. I don't know yet, but sustainable planning, small-town planning, has been impressed upon us as desirable by the corporate actors as well as people who work in the planning offices in the small towns. They—we—need that type of intelligence. Maybe in part to counter the pressures of a corporation, but equally, to try to think on the greater behalf of the small-town citizens.

Notes

1. The 2022 census listed Bentonville's population at 57,936, while in 1990 it was just 11,257 (U.S. Census Bureau).
2. The company has long been spelled "Wal-Mart Stores, Inc." But in 2018, it officially changed the name and spelling to "Walmart, Inc." (Walmart, 2022).
3. In a conversation with the authors, 2020.
4. In a conversation with the authors, 2020.
5. 2020 population (U.S. Census Bureau).
6. The Transportation Investment Generating Economic Recovery, or TIGER, grants are discretionary programs from the U.S. Department of Transportation for road, rail, port and other transit projects designed to address national objectives.

References

Annis, R. (2020, November 20). Northwest Arkansas Is Offering You a Free Bike and $10,000 to Move There. *Bicycling*. Retrieved from www.bicycling.com/culture/a34738872/northwest-arkansas-life-works-here-moving-initiative-bike/

Bonacich, E. and Hardie, K. (2006). Wal-Mart and the Logistics Revolution. In N. Lichtenstein, *Wal-Mart: The Face of Twenty-First-Century Capitalism*. New York: The New Press.

Gamble, D. and Heyda, P. (2016). *Rebuilding the American City*. London and New York: Routledge.

Gilligan, C. (2023, May 3). The States With the Highest Poverty Rates. *U.S. News & World Report*. Retrieved from www.usnews.com/news/best-states/slideshows/us-states-with-the-highest-poverty-rates?slide=11

Golden, A. (2022, October 28). NWA's Boom Demands Bold Planning, Experts Say. *Axios NW Arkansas*. Retrieved from www.axios.com/local/nw-arkansas/2022/10/28/nwas-economy-boom-bold-planning

Northwest Arkansas Design Excellence Program (2024). Retrieved from https://www.waltonfamilyfoundation.org/grants/northwest-arkansas-design-excellence

NWARPC (n.d.). Retrieved from www.nwarpc.org/razorback-regional-greenway

Repovich, M. (2023, October 8). P. Heyda, interviewer.

Rowe, P. (2009). The Spatial Transformation of American Cities. In J. Tatom and J. Stauber, *Making the Metropolitan Landscape: Standing Firm on Middle Ground*. London: Routledge.

Souza, K. (2016, February 22). Northwest Arkansas Tourism and Travel Sector Ends with Record Results. *Talk Business & Politics*. Retrieved from https://talkbusiness.net/2016/02/northwest-arkansas-tourism-and-travel-sector-ends-2015-with-record-results/

Sparkman, W. (2023, June 16). Northwest Arkansas' Bet on Cycling is Paying Off. *Axios Northwest Arkansas*. Retrieved from www.axios.com/local/nw-arkansas/2023/06/16/northwest-arkansas-cycling-industry-economic-impact

University of Arkansas Community Design Center and Washington University in St. Louis Sam Fox School of Design & Visual Arts (2007). *The Rail Transit Design Studio: Visioning Rail Transit in Northwest Arkansas: Lifestyles and Ecologies*. University of Arkansas Community Design Center.

Walmart (2022). From Humble Beginnings. To Redefining Retail. Retrieved from https://corporate.walmart.com/about/history

Walmart Corporation. (2023). New Home Office: Regeneration: Breaking-Ground. Retrieved from https://corporate.walmart.com/about/newhomeoffice/breaking-ground/regeneration

Walton Family Foundation (2022a). About Us. Retrieved from www.waltonfamilyfoundation.org/about-us

Walton Family Foundation (2022b). Walton Family Foundation. Retrieved from www.waltonfamilyfoundation.org/our-work

Section 2 **Satellites**

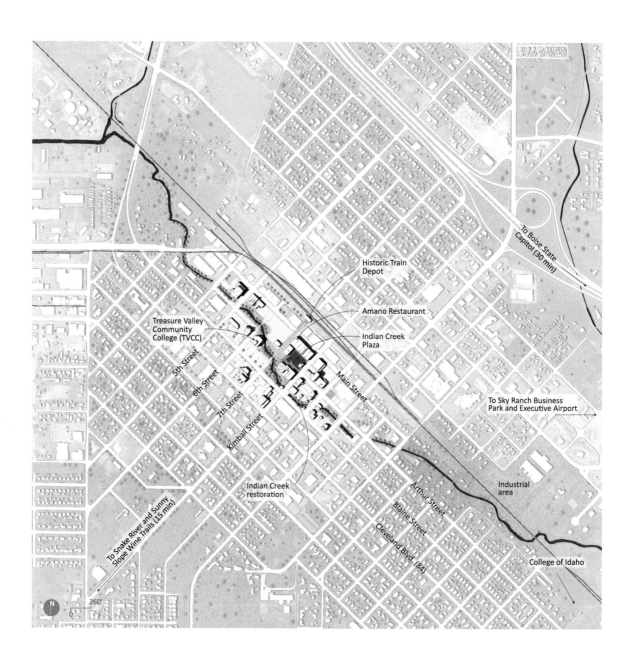

2.1 Caldwell, ID

Tucked up against the border of Oregon, Caldwell is located in picturesque Canyon County in Southwest Idaho. Known as the Intermountain West, it's one of the fastest growing regions in the United States with transplants flocking there for greater affordability and new business and real estate markets. Idaho is a state built on small communities, but Boise is the state's capital located just 28 miles (45 km) east of Caldwell. Boise is one of the fastest growing cities in the state and, as a percentage of the population, Idaho's growth rate ranks near the top in the United States (3 percent from 2020–2021). That puts Caldwell in the orbit of one of the fastest growing cities in one of the fastest growing states. As the valley matures and land prices escalate, satellite towns in the metro Boise area find themselves in the crosshairs of increasing development pressures. Caldwell is one such place, with a population that has grown 137 percent since 2000.[1]

When the Idaho & Oregon Land Improvement Company filed a plat plan for a new town in 1883, they oriented streets parallel to the Oregon Short Line Railroad tracks along Front Street. These tracks linked Wyoming to Oregon. Front Street has since been renamed Main Street and, like many small American downtowns, it is lined with an eclectic assortment of historic buildings that share party-walls and are built up to the sidewalk. The condition of the Romanesque, Renaissance, Spanish Colonial Revival and Art Deco facades vary and closely track the town's economic ebbs and flows. Today, following generations of neglect and a long incubation time, Caldwell's Main Street and surrounding downtown blocks are incrementally filling back up with new tenants and activities.

The catalyst for the downtown's revitalization had to do with public-sector investments in new downtown spaces and in an anchor institution. But investments were also made in something that was already there, but hidden. The "daylighting" of downtown Caldwell's Indian Creek uncovered a waterway that had been completely buried in a culvert below buildings, streets and surface parking lots since the late 1950s.[2] Completely out of view, many people had forgotten the creek was even there. Since the 2009 rediscovery and reinvestment in this resource, the downtown's tight street grid has become interwoven with a sinuous band of open space that meanders a block or two away from Main Street—and catalyzes further investments between them.

In the second half of the 20th century, the downtown's fortunes had declined and commercial activity retreated to the nearby mall. Vacancy and crime increased as the core languished. However, in the early 2000s municipal

In a planning and construction effort spanning a decade, Indian Creek was removed from its culvert and naturalized through downtown Caldwell, Idaho.
Credit: Courtesy of Destination Caldwell.

leaders established an Urban Renewal Agency and initiated a downtown master plan. Efforts were led by a six-term mayor who had cultivated over twenty years of trust, providing the benefit of having and acting on a long-political-view, something many cities and small downs don't experience when administrations change every four years. The master plan framework's most visible physical intervention resulted in the ambitious and protracted effort to open the waterway back up. Despite significant funding setbacks, the infrastructure project was one of many initiatives that was thought could actually improve the regulatory, economic and physical condition of the struggling downtown. **Today, Indian Creek and its adjacent greenway have formed the backbone of an economic and environmental revitalization effort that stabilizes Caldwell by re-centering its downtown**.

Small towns in beautiful settings have become attractive get-away locations for outdoor enthusiasts on a budget or for people looking to avoid the usual overcrowded destinations. As mentioned in the book's introduction, this is also true for people who became more flexible during the COVID-19 era to move to more affordable towns and work remotely. Prior to opening up the waterway, downtown Caldwell's slow revival was supported by years of public policy that prioritized families and sought to overcome perceptions that downtown was unsafe.

On the western edge of Caldwell, Indian Creek flows into the Boise River. The waterway irrigates the valley and supports a proliferating wine industry that is located on the same latitude as the south of France. Caldwell is gateway to the Sunnyslope Wine Trail within the Snake River Valley, an American Viticultural Area (AVA) that has more than a dozen wineries and three times as many vineyards within a thirty-minute drive of Main Street downtown. The overall geography covers five million acres along the Snake encompassing more than 90 percent of Idaho's vineyards. The valley has contributed to Idaho's $3 billion industry that continues to expand.

Despite grapes and growing tourism, "Big Ag" is still the top industry in Idaho, a state with more cattle than people.[3] This is true in Caldwell, too, where commercial agriculture remains dominant. The Treasure Valley landscape is comprised predominantly of family-run farms larger than 1,000 acres that utilize high-tech contemporary farming processes to increase efficiencies and production. However, like much of the agricultural industry across the United States, these companies rely on a substantial influx of permanent and migrant workers. Caldwell's proximity to these farms, and its relative affordability compared to other towns in Canyon County, makes it an attractive location for farm laborers, many of whom are of Mexican or Latin American origin.[4] Among permanent residents, the City of Caldwell is nearly 40 percent Hispanic in a state that is over 90 percent white, resulting in both tensions and collaborations. The spatial distribution of residents in Caldwell still reflects racial/ethnic divisions along neighborhood lines, but the center of town where the creek was restored is where two neighborhoods—one majority Hispanic and the other majority white—meet, along Main Street (Race, Diversity, and Ethnicity in Caldwell, ID, 2024).

Water Works

"Daylighting" projects involve liberating waterways that are buried in culverts or pipes, covered by decks or otherwise removed from view (Pinkham, 2000). The efforts vary greatly in the degree to which they re-naturalize a waterway, but generally, more urbanized locations are severely constrained as they have to negotiate private property, existing roads and infrastructure.

The challenges are many. For one, it is expensive to purchase valuable downtown real estate, to move or remove commercial buildings and other components of built infrastructure, and to engineer the design and construction of a daylit stream to meet flood control parameters, water quality and habitat needs in a downtown setting (Hoobyar, 2002). Excavation efforts require regrading to address channel alignments and geometries. Pulling up a culvert and creating a new landscape involves a significant amount of earth moving, including potentially contaminated soils that need to be removed or capped. These are long-term, protracted efforts that take time, usually transcending political terms and requiring strong public education and outreach to build consensus and generate support. Moreover, the daylighting of rivers or streams requires working beyond the scale of the waterway itself to integrate with the overall urban storm-water management system of the watershed. The complexities and costs surrounding the process of excavation, particularly in downtowns, explain why so few communities have been actually able to achieve it.[5]

View of the newly restored Indian Creek and pedestrian infrastructure in downtown Caldwell, Idaho (2020).

Nevertheless, the environmental benefits of daylighting are considerable. It increases the hydraulic capacity of the waterway, improves water quality and reduces harmful runoff and heat-island effects. Daylit creeks restore ecologies by establishing new open space corridors in downtown settings, recreating aquatic and riparian habitats and providing recreational amenities for the community. While costly, daylighting may actually prove to be less expensive than replacing a deteriorating culvert below grade. In Caldwell, the process took seven years to complete. The project spanned a twelve-block area downtown, adding six acres (2.4 ha) of new open space and 3,000 feet (900 m) of pedestrian and bike trails. The initiative galvanized community spirit and laid the foundation for future urban redevelopment.

To enable implementation, Garret Nancolas, the mayor at the time, formed a core area steering committee of residents and business owners that collaborated with the Army Corps of Engineers who manage the waterway. The Corps suggested the use of Section 206 of the 1991 Water Industry Act for funding the project since it authorized them to undertake work focusing on increasing aquatic habitat. Utilizing these federal funds also allowed the cost to be shared, with the federal government picking up 65 percent and the local government responsible for the remaining 35 percent. More than a dozen other agencies were involved in the cross-sectoral endeavor, including the Idaho Transportation Department (who funded the trail system), the Federal Emergency Management Agency (FEMA), the Environmental

Protection Agency (EPA), Department of Transportation (DOT) and the National Park Service which addressed parts of the creek design and trail system master plan.

Major infrastructure investments are protracted efforts, even though small towns boast the ability to streamline private development. Elements of Caldwell's downtown initiative took time, while other related developments could be moved through the approvals at "unheard of" time efficiencies, drawing on the "Red Carpet Team" set up to attract developments more generally. Mayor Nancolas describes that strategy:

> We know that time is money. We know that getting permits issued is money. We know that process is frustration if it's not smooth. So we developed a Red Carpet Team. We promised that we would get your building permit issued in ten days. That's unheard of in the manufacturing and commercial sectors. When the Fresca Company came here, it was a 190,000 square foot building. We issued their building permit in eight days. That's unheard of. The theory of the Red Carpet Team was, let us help you from the very beginning. As soon as you're thinking about a project, you come to us, we'll pull together engineering, public works, water, sewer, fire, whatever you have to have in your permitting process. You tell us what you want, we'll give you every bit of ammunition and all the information you need. We'll give you a checklist and we will walk you through it. We had one person assigned to each project so that they would hold their hand and make sure they had someone with them. And it worked.
>
> (Nancolas, 2020)

It worked, but not without significant advocacy required to generate public support and political buy-in prior to formalizing the intergovernmental partnerships. Not insignificantly, Caldwell Mayor Nancolas would go on to serve six terms as mayor. He spent over a decade building connections needed for the creek restoration, a process of relationship building that he distinguishes from partnerships.

> Relationships, to me, lead to partnerships. And you can't have a partnership unless you trust each other first with a relationship of understanding and common goals. Whether you agree all the time or not is irrelevant if you have a relationship. If you don't have a relationship, a disagreement can blow everything up.
>
> (Nancolas, 2020)

The partnerships were working well, until they weren't. The federal government stopped funding Section 206 shortly after Caldwell's daylighting project began. By this time, the city was already well underway with site acquisition, building demolition and design of the creek restoration. It caused them to stop work for more than a year, re-evaluate the project and form or strengthen other partnerships to support and fund the effort. To the detriment of workers, the funding cut freed the town from meeting the prevailing wages required for the use of federal funds. The labor cost for the project was cut in half. Other changes also emerged that enabled the right-of-way to do more than just address riparian habitat—another requirement for the funding.

Residents dance at a planned event and farmers' market called "Tuesdays on the Creek" at Indian Creek Plaza, in Caldwell, Idaho. Credit: © Eleven One Photography.

Eventually, the city acquired twenty sites along the 1,550 feet (472m) of Indian Creek, hop-scotching around property owners that wouldn't sell, as the mayor was determined not to use eminent domain to achieve the objective. The naturalized greenway is now the centerpiece of a robust calendar of outdoor events, anchored by two major festivals. Some argue that "public programming" is an over-curated attempt by cities to control sanitized versions of "publicness." Others applaud events like the Indian Creek Festival that bring tens of thousands of people to the once covered-over waterway for kayak races, dancing, live music and art in the late summer or for the popular winter wonderland lighting spectacle toward the end of the year.

Nourishing/Culture

Canyon County is a major agricultural area in the United States and one of the largest firms is headquartered in a contemporary campus in downtown Boise. The J.R. Simplot Company is considered Idaho economic royalty, a $6 billion food and agribusiness with 13,000 employees. (Simplot pioneered frozen french fries and distributes them to McDonald's and Burger King franchises around the world.) They have a large local footprint in Caldwell with over 1,200 employees. Caldwell is also headquarters of Fresca, one of the largest manufacturers of tortillas; and the Crookham Company, one of the largest seed banks in the world. Although the immense scale of these industrial complexes is at odds with the small town in which they operate, the companies form a substantial employment base that many communities envy.

The oversized corporate presence in manufacturing is complemented with small businesses that advance a local, culinary experience. Salvador and Becca Alamilla own a restaurant in downtown Caldwell called Amano. It is located on a prominent corner along Main and 7th Streets in a former bank building with a hand-carved wooden ceiling and huge windows. The restaurant is a destination food "experience," drawing customers from Boise and beyond for its made-from-scratch tortillas and Mexican dishes. Their mix of customers speaks to the profile of the town: "Our demographic isn't necessarily the Latino population," Sal says.

> But we get the sons and daughters of the Latino generation: people in their 20s and 30s that are coming here that are bringing their parents. But we're not necessarily getting the parents right off the bat. Our demographic for lunch is mainly composed of white, 40-to-50-year olds and our weekend crowd is kind of more mixed, so about a 50/50 mix of the Latino and the white population.
>
> (Alamilla and Alamilla, 2020)

An American Town

Ninety-three percent of Caldwell residents are U.S. citizens (World Population Review, 2024). Compared to Idaho's population where 13 percent of people have Hispanic origin, Caldwell's residents are 38 percent Hispanic (United States Census Bureau, 2023). This number does not account for the substantial transient Mexican-American laborers that reside in the area during the growing seasons and wouldn't be counted in the census. The population skews young, where the median age is 29.5 years old, perhaps influenced by the College of Idaho and Community College campuses nearby. As the Hispanic population grows in Caldwell and surrounding towns, economic prosperity grows with it. In 2024, Caldwell's poverty rate was at 13 percent of households living below the poverty line (World Population Review, 2024). This was the same as the U.S.

Cars with character serve as emblems marking old and new times in the public spaces of Caldwell, Idaho.
Credit: (left) © Eleven One Photography; (right) Authors.

average (13 percent), but the most common ethnic group living in poverty in Caldwell was white.[6] Caldwell represents an American multicultural town, reflective of national trends toward upward mobility for a richer, more diverse multiracial society.[7]

Satellite Connections

Not all growth is smart growth. There is significant sprawl along Interstate 84 between Caldwell and Boise and the suburban development patterns are cannibalizing the agricultural fields and bucolic landscapes that attracted people here in the first place. Heavy traffic along the primary highway corridor is frequently congested, and there are few mechanisms to fund urgently needed public transit in Idaho, despite the network of legacy freight rail lines that slice across the landscape and facilitated the region's expansion.

Like the Jamestown, New York case in this book, the initial growth of Caldwell was directly tied to a network of railroads that served the town, yet the historic train depots' roles have since changed. Caldwell's iconic Main Street depot maintains a central place in the downtown, on axis with one of the main bridges that crosses the creek. The building was purchased in 1989 by the city from the Union Pacific Railroad and is listed on the National Register of Historic Places. It opened in 2009 as an event and interpretative center which can be rented out for large gatherings—although arguably could still benefit from other uses and more vitality. Freight trains still rumble through the right-of-way on a daily basis, but there is limited bus transit in Caldwell and no passenger or light rail. One can hope that the public spirit of the creek restoration project might spread confidence into other important public transit infrastructure projects, especially as urgencies of global climate warming demand transitions from fossil-fuel based forms of private mobility. For now, the state legislative political structure prevents the town from acting autonomously on those kinds of initiatives. Developer Skip Oppenheimer details this conundrum:

> Unlike most states, we're one of a handful that don't have local option tax. So to fund transit is very difficult because we have no real way to vote on whether we want to tax ourselves or not. It's our legislature that's pretty far off on one edge. They are afraid that if they turn over the ability to vote on local option, people will vote for taxation. And "taxes are bad." We've got to figure this out because of sprawl. If you want to continue to have a vital downtown and attract employers to downtown, you've got to have some way to get people in and out. And we don't have a very good transit system. We don't have a real way to fund a really good transit system. And you could see nodes, I mean, you have Caldwell, Nampa, Meridian, Boise. … You could build around those downtown cores and have enough concentration of activity that you could actually make some form of transit viable. Nampa was once a big railroad center.
>
> (Oppenheimer, 2020)

The historic train depot is framed at the end of the bridge alignment across Indian Creek in Caldwell, Idaho. The station was restored in 2002 and now functions as an interpretive and event center that can be rented out to the public.

2.1 Caldwell, ID

Mid-block but front and center, Indian Creek Plaza forms the new hub of communal life downtown one block removed from both Indian Creek and Main Street.
Credit: © Eleven One Photography/Destination Caldwell.

Flexibility of Being Small

Indian Creek Plaza is Caldwell's popular new public square where activities convene downtown. The square is offset one block from both Main Street and the Indian Creek corridor, so it was originally an unlikely place to center urban public life. The location had not been directly referenced in the master plan but it emerged through a series of fortuitous alignments that caused leaders to take note and shift strategy as other planning was happening. A dilapidated King's store became available for purchase in that location, and leaders worked quickly to get site control. The building's eventual acquisition and subsequent demolition created a new void in the downtown fabric equally distant from both the creek and Main Street. While the urban design plan had always envisioned a public plaza downtown, it was anticipated in another location adjacent to Indian Creek and on-axis with the historic train depot. The plaza was to be the center point of a cultural center anchored by a new city hall that would consolidate municipal offices in one building and include a variety of retail and office spaces to connect the creek, streets and pedestrian pathways with this civic plaza. However, when financing for the city hall complex became unfeasible as a result of the national 2009 recession, the area remained vacant and this public space emerged as a compelling alternative.

The most effective design plans allow for some degree of flexibility or modification as circumstances, economics or individuals change. But certainly, being a smaller sized community enabled city leaders to pivot from the plan via direct connections to the particularities and actors involved in the given situation. The vitality of Indian Creek Plaza today can be attributed to its spatial definition and design flexibility as well as its programming. Interestingly, the current site of the plaza has proven to be in a better location than initially proposed in the master plan because it is surrounded by historic buildings that create urban enclosure. And arguably the location off-axis *expands* the pedestrian network, even without the civic anchor of the proposed city hall next door. Caldwell's downtown Treasure Valley Community College building is also a factor and an effective proxy for a municipal complex. When it was completed, it brought 40,000 square feet of new construction and students to a prominent bend of the reclaimed creek downtown. The institution is complemented by the College of Idaho, another much larger campus less than a mile away. The College of Idaho is the state's oldest private, liberal arts college.

Town and Country

As urban theorist Neil Brenner writes, downtowns today are not entirely self-sufficient, but are increasingly reliant on intensive ever-expanding processes of urbanization in and beyond the periphery (Brenner, 2014). The revitalization of Caldwell's downtown core is supported, in part, through the establishment of an industrial zoned area just outside the town center. This area is called Sky Ranch Business Center. It is a designated Foreign-Trade Zone 280, and comprises 300 acres of commercial and industrial uses with easy access to Interstate 84, Highways 20/26 and the Caldwell Industrial Airport. The business and logistics center has added manufacturing jobs, and its economic success led to another industrial park nearing completion called North Ranch Business Park. That 120-acre property will accommodate over 1.4 million new square feet of industrial flex space.

From a narrow economic perspective, the downtown investments and the development on the outskirts are synergistic. By leveraging the large industrial tracts on the periphery, Caldwell was able to finance the downtown improvements. At the same time, such massive logistics landscapes typically harm their regional communities in the long run, as they require new infrastructure, they amplify environmental degradation of air and water, they increase emissions and heat island effects and they contribute to the erosion of urban space and design—alongside compromises to agricultural areas and species diversity responsible for the health and beauty of the region. The strategy of supporting a small town with unhospitable industrial landscapes shares similar risks as with larger cities and industries reliant on extraction elsewhere. These intractable relationships across scales of urban settlement perhaps speak more to the larger growth-oriented political structures in the United States that enable unregulated resource exploitation in planning.[8] Nonetheless, such tensions are more acutely felt at the smaller urban scales.

Rebuilding an American town in Idaho is a multicultural project. It brings together large-scale issues of industrial labor and expansive land uses outside of town with evolving ideas of how downtowns and urban design can unite the community. As once-vacant Main Street stores are gaining traction with new commercial tenants, the downtown celebrates together at events near the restored creek. Some large surface parking lots do remain and provide the ground for future infill development or creative open space strategies. The locally restored environmental and social ecosystem downtown models a microcosm of new priorities for the region's broader landscape, waterways, business health, walkability and social life.

Perspectives

Mayor Garret Nancolas, Mayor of Caldwell from 1998 to 2022

Caldwell has always been an amazing community. I've been mayor for twenty-three years, and I was on the Council for eight years prior to that. It's like trying to herd cats sometimes. But if you don't, you lose momentum in one area that could affect the others.

By the mid-1970s or so, Karcher Mall (in the suburbs) had come, and the stores that we loved all went there. Downtown Caldwell began this downward spiral. We went from a robust, movie-theater, no-crime, every-storefront-filled, to a 65–70 percent vacancy rate. I literally could've sighted my rifle downtown and not had to worry about hitting anybody. I always had this mystery of, how do we bring this back?

In 1999, I was this new, naïve Mayor. Cliff Long, who was the State Department of Commerce Director, said, "Do you have an Urban Renewal Agency?" I said, "What's that?" Of course, he educated me on what that was, and I came back to our City Finance Director whose response was the same thing, "Well what's that?" It allowed us to take all this discussion, all the interest, all the concerns and create a firm fifteen-year plan known as the Caldwell East Urban Renewal Plan. We spent a year drafting it, with town hall meetings,

A view of Caldwell, Idaho in the 1960s.
Credit: Historic Caldwell/City of Caldwell.

meetings with other affected tax entities, with the school districts, with the college, with the business community, the chamber of commerce and the state. The Plan addressed everything: water, sewer, roads, bridges, economic development, parks, recreation and everything in-between. We made it big and inclusive, because I've always believed that the fear is not that you set your goals too high and you don't achieve them, it's that your goals are too low and you *do* achieve them.

. . .

There was a meat packing plant in Nampa [the next-door town] that dumped their blood and guts into the Indian Creek. All these older homes had pipes that spewed raw sewage into the creek. As a matter of fact, if you go down the creek, you can still see some of the old sewer pipes. So, in the 1950s they covered the creek up to deal with that issue because it was unsightly, it was smelly. That was the downtown that I remember as a kid; it was all underground.

We had the good fortune to run into Debby and Brayton Willis, from the Army Corps of Engineers, and started talking about applying for a grant to daylight this creek. That was going to be the centerpiece of our downtown revitalization. The grant was a 65/35 match; the Army Corps paid 65 percent and we paid 35. So we start buying the properties and beginning the design stages, and then Congress pulled the grants. We were already in for probably $2 million dollars with buying property, design work, and everything else.

So we re-wrote the Urban Renewal Plan and decided to fund the project ourselves. We invested seven million dollars in this renovation of the creek, which took us seven years. In an old downtown like ours, you have a gas tank or a dry cleaner on every corner. So EPA got involved, and I learned more about phytoremediation, and perc and petroleum more than I ever wanted to know.

When we started this project, we tried to involve every entity because we wanted the creek to have fish in it. We probably had twenty, twenty-five partners that helped us get through it. Fortunately, one of our city council members was an environmental specialist. Some property owners said they would sell, then said they wouldn't, so we had to redesign the creek because we refused to use eminent domain. We wanted it to be a cooperative project. Of course, the sewer and water issues caused us to take a look at the dilapidated sewer system in downtown, which we then had to completely redo. For the business owners downtown, that was hard to deal with, and yet they stuck with us.

And then, in 2008, the economy went in the toilet. We lost a third of our assessed property value in one year. But from 2003–2006 when things were exploding, the Urban Renewal Agency did well, so thank heavens we still had some resources. While prices were low we put in all kinds of infrastructure in the downtown area. We started building our industrial park out in Sky Ranch Business Park. We could get work done extremely cost effectively because people were hungry. We had to create jobs because Caldwell's average household income was literally at the poverty level. We didn't have the jobs, we didn't have the downtown, we didn't have the manufacturing. We had farm jobs, which are great, but we wanted basic-sector jobs.

. . .

Over a million holiday lights illuminate Caldwell's annual Winter Wonderland festival at the restored creek.
Credit: Marilyn Isaac Photography.

When we started building the creek we wanted it to be a destination point. We wanted it to be beautiful, a place for families, and we wanted to create some critical mass that would help the businesses. But if we hadn't addressed crime it wouldn't have made any difference. We wanted to light every rock, every tree, every bush, every bridge—lights everywhere, so that people could enjoy the atmosphere downtown. We connected all the street lamps so they could have music playing. And people started coming downtown and seeing the lights. We have pedestrian counters on this little bridge over here on 7th Street, where we had about 170,000 crossings for our winter wonderland, between the middle of November and the middle of January. We started with 600,000 lights, we're at about a million lights in downtown now. It's all free. We had 15,000 people show up at the creek lighting this year. 15,000 people in downtown Caldwell! That's huge, right?

. . .

When we talked about daylighting the creek, a guy stood up in a town hall meeting and said, "This is a waste of money and time, we're not gonna stand for this, and if you go forward, we'll recall you." That's kind of scary. But we knew it was the right thing to do. And once it started coming together, then those naysayers became some of the biggest cheerleaders. Matter of fact, the guy that stood up and threatened the recall election actually called me years after the fact and said, "I'm sorry, I was wrong, this is amazing."

. . .

Part of our framework masterplan was a community or civic space. The next thing you know the King's store downtown closed. They were an old department store, kind of like a five-and-dime deal. They had deeded the building to the Salvation Army and we heard the rumor that they were going to put it on the market. So, we convinced the Urban Renewal Agency to buy the King's building, and that became the center of our goals to create this plaza.

The costs kept escalating. We thought it was a five million dollar project, we're in at over seven million dollars. So, we told Destination Caldwell and the business community to get the merchants to sign a Business Improvement District (BID) petition. Because we can't just fund this out of city revenue. It has to have buy-in, it's an ongoing expense, and when you start talking about general fund dollars, now you're competing with police officers and firefighters. The BID generates about $270,000 a year, and we signed a contract with Destination Caldwell to operate the plaza. They do all kinds of things that cities aren't professional at, like they're great at marketing, and having all these special events that obviously help fund their budget.

. . .

We now have six acres of park in downtown Caldwell, which is much better than asphalt for the ambient temperature, for trees, and just aesthetically. We also put in incentives for LEED-certified buildings[9] in our codes, and we have encouraged porous asphalt. I think we were one of the first cities to jump on adding LEED certification to our codes and incentives. We didn't want just another downtown where you have building upon building, and no open space and no trees. We wanted to use the environment as well as business to create the gathering place and the destination point. And the creek became the focal piece of that.

. . .

This is still Idaho. Private property rights are a huge issue in all areas, but particularly here. We had people wanting to hop-scotch out into the county, but we didn't want them to develop without having water and sewer and walking paths and so on. So we allowed them to have extra-territorial agreements, to encourage them to connect to water and sewer, so that we could protect the environment. It did create a little bit of that sprawl, but on the reverse of that we created incentives for infill development in our codes and our ordinances. Some of these are starting to come back in through PUDs (Planned Unit Developments), or through allowances for narrower streets, so that we can get infill and stop some of this sprawl. It's just been an interesting political dynamic trying to deal with the rules from the legislature plus what our goals were as a community.

Keri Smith-Sigman, Former CEO, Destination Caldwell

Destination Caldwell is an economic development non-profit. We started the AgVenture Trail. While we do strive to bring in tourism and make Caldwell known for our agricultural heritage, it's how we're doing it that is really unique.

Canyon County is *big farm*. I think we're number three in the world for seed production. When I was speaking to local groups, most of the time it was big farm at the table. If a home owner in a subdivision decides they're going to plant corn for their kids and there is a seed crop within one mile, that seed crop is wasted because of cross pollination. We lost a couple of seed companies because of the growth in Canyon County.

We also have smaller farms like Babby Farms, which is like an exotic petting zoo, and Valli-Hi Angus Ranch. We have McIntyre Farms—completely sustainable farms with grass-fed beef and chickens. We were talking about how to get more people to come and buy directly. I said, I have a whole bunch of city slickers who would love to have an experience to come and feed your chickens. Will you let them? We started building those relationships.

The growing season closes right around Thanksgiving so I need to get people to start putting in greenhouses to have produce in the winter. We just don't have that consensus yet. But Sunnyslope Wine Trail is open year-round. Right now, we're not doing any paid advertising for AgVenture, but once it hits spring, we'll do that. We also are doing the super cool Farm to Fork Farmers' Market and we do a summer concert series every Tuesday night, the middle of May through middle of November. The ice rink goes from middle of November until February.

. . .

As soon as Indian Plaza opened, I got more calls than I could put people in buildings. When you see empty buildings it's not that I can't fill them. It's that almost 50 percent of these buildings, including the one we're sitting in, is owned by the Jenkins family, Bob and Kelly. They're local. Ten or 15 years ago when these buildings were going up for sale we had like a 90 percent vacancy rate. There was a porn store, there was an X rated movie theater. Just junk. Bob started buying these buildings, one building at a time, with an old school attitude of "I'm not going to go into debt for this project; I'm going to methodically work on every building." He took advantage of our facade improvement program for what was probably the ugliest building on the plaza—a 35 percent match on their investment. My job was to go out to each of these property owners on the plaza and get them to participate.

The city owns the land and we are contracted to manage it and do 150 events per year. When I was doing my sales job, I told my consultant we'd actually do 250 because that's what Rapid City had done. He said "Don't do 250. It'll burn your staff out." So, when I went to the city for the contract, we negotiated down to 150 because I wanted to exceed expectations. This year

Now in its twentieth year, the September cardboard kayak race is a major fundraiser and popular annual event on Indian Creek.
Credit: © Eleven One Photography.

we did 315. That includes ice skating days. We get a lot of traction on our large games that we put outside like ping-pong and cornhole. One of the struggles that I have is getting people to stay open later. When I do recruitment, I talk to them about business hours and the synergy that we have to create it. It has to be a little bit of a mall mentality downtown because we all need to be open the same hours.

Our main school districts are Valley View and Caldwell. In both, white residents are the minority. I just love that we have pulled another demographic into the downtown. We had a big Cinco de Mayo partner event with Rancho Radio. It was the first time I saw the perfect blend of Caucasian and Mexican culture. They were all having a good time, and it was heartwarming. The same thing with ice skating. It doesn't matter who you are. Everyone comes and skates.

Vanessa Fry, Research Director, Idaho Policy Institute, Boise State University

Caldwell wasn't originally a bedroom community for Boise, it was more of an Ag base of the farmers, ranchers, agricultural workers—there wasn't a ton of commuting. I think Caldwell leaders started to realize that they're becoming a more urbanized area because they are proximate enough to Boise that people are living there and commuting. It only takes a half hour to drive there without traffic.

When you head west from Boise you have highway 84, Eagle and Meridian, and then Nampa and Caldwell. Those all used be very rural communities and they're becoming more urbanized. Caldwell also has the College of Idaho. It's the only liberal arts college in the state. Even though it's

A vacant block near Main Street in downtown Caldwell in 2020 that has since been redeveloped with a hotel.

a fairly conservative community, having that liberal arts component helps them think out of the box a bit. When they were in the planning phases for daylighting the creek, people were so nervous, like, "Is this going to work? Are we actually going to see more people downtown along the river?" Certain city staff recognized the aesthetic benefits as well as the potential for economic development. Because it's such a rural state, our Department of Labor and Department of Commerce have put a lot of time and effort into these communities. It was actually the Idaho Transportation Department, who had some funding for pathways, and Idaho Smart Growth and some other local organizations that helped communities to consider revitalizing their downtowns to be more walkable.

A Conversation

Becca Alamilla (BA) and **Salvador Alamilla** (SA), Amano Restaurant owners

SA: We got really lucky with this space. It used to be a bank in the late 1800s and early 1900s, and it's been a lot of things since then, but over the years downtown Caldwell started to become a ghost town. But now the businesses are coming here, even businesses from Boise that are looking for opportunities because this seems like a new market. There's a lot of new people moving to the area.

We make our tortillas from scratch, meaning that every day we make nixtamal, which is corn and calle, and we cook it and then grind it. The corn is from Oaxaca. We make a lot of things in-house. That's the foundation of the restaurant. When we first started it brought many of our team members a lot of memories—the tortilla just smells and tastes different than if you were to get store-bought. That's what I wanted to present here in Caldwell. It was a different approach from what people are used to, although these are old family recipes. We're introducing something old, but for a lot of people it's something new that they haven't really had before.

. . .

We were looking for a space and every time we would find one, there was always something that came with it. Either we couldn't negotiate the price per square foot or there were things the landlord wanted that we couldn't afford. We lost at least two deals before we found this space. We knew that we needed to be in Caldwell and we knew that downtown was the place to be. On the weekends there's thousands of people here. And we're right on the corner.

The other day I picked up some hinges for the doors, and the lady there told me that she wasn't too happy with the changes that were happening, that it brought in a lot of new people and people from Boise. For her the impact of this was a negative. But for a lot of other people, it's positive. We ran into a lot of backlash since we're a Mexican restaurant. We got, "Hey does this community really need another Mexican restaurant? What's different between your restaurant and the next Mexican restaurant across the street?" There was a lot of that.

Amano Restaurant, authentic Mexican cuisine occupying a former bank building in downtown Caldwell, Idaho.
Credit: (left) © Eleven One Photography; (right) Authors.

BA: I think he's even understating it, to be honest with you. We had a prominent wine-owner pull us aside and say, "You guys really aren't welcome. There's so many Mexican restaurants here, unless you are gonna be the best in the whole Treasure Valley, we're just not interested in having you." And we were like, "Well, that's our goal. We don't go into anything trying to be 100th on a list somewhere, you know?" That was pretty surprising, I think, for us, maybe because of our backgrounds. Sal doesn't have a culinary degree, but he's been an executive chef at several places. I've been a part of several startup organizations, one of them through Teach for America. I started a whole charter school. But I think because of that there was a little bit of doubt, like "Oh they're just young and naive."

SA: It's a family business even though it's a grand and beautiful space. My mom works here, my dad works here, my brother works here. We did run into a lot of difficult situations and conversations about leaving. The idea was to talk to community leaders, because it's a hot market here, and community leaders, especially people that sit on the board of Destination Caldwell, have a lot of say as to who gets to be here and who doesn't.

BA: The public sector definitely got the Indian Creek Plaza rolling. It was a dream of a couple of community members, and honestly, if the plaza wasn't here we wouldn't have taken the risk because it was a ghost town here just two years ago. I went to the College of Idaho, which is a mile and a half up the road, and the joke was to not come down here because you're gonna get shot, because of those abandoned buildings. I think that was the reality in the 1990s and they said in the 1970s it was like that too, so it kind of went in waves.

SA: Caldwell developed the name of "a place not to visit." And to this day, the people who haven't been out here still question it.

. . .

BA: We always felt like our food would be a destination for people, kind of like experiencing going to Mexico, but locally. We started branding some of our stuff using imagery from our trips to Oaxaca to provide the foundation that we're connected to that, and it kind of had this "cool-kids" effect. People from Boise started coming here, and all of these food bloggers started writing great things. Within our first two months we had several newspaper articles by people who are pretty notable in the food scene here. We were very fortunate that we so quickly got their support. Like 25 percent of the people who come in here every day are taking a forty-five-minute trip from Boise. If we were to rely just on the population here in Caldwell we couldn't make it. We weren't surprised by that though because in our demographic study before we opened, Boise was our biggest demographic followed by Caldwell. We knew we would have people coming here from Boise, or hoped, at least.

. . .

SA: We can fit about 120 people, open seven days a week. Saturday and Sunday we open for brunch, so we have chilaquiles, huevos divorciados a menudo, all your Mexican classics. It's a big commitment to be open seven days a week. I think the easier thing for us would have been to close a day or two. But the community needs spaces like this, where they feel welcome, where they come in and stay warm and have a great meal. And it should be offered more than five days a week. In Caldwell there's not a lot of places that are open on Sundays.

BA: That's one of our best days because everywhere else is closed.

Jim Everett, Co-President, College of Idaho

It was pretty cool to watch the community rally around the creek revitalization. They had really been beaten down; they were second fiddle to Nampa and Canyon County and way down the list from Boise and just had a self-esteem issue. To take on something that even the professionals that do this all the time were doubtful about really created confidence.

Urban renewal in Idaho is extremely controversial. The mayor has been able to do this in a way that is just masterful. He's been able to unite people in a way that other communities have tried to do and just can't. I'm originally from Michigan and I love Idaho, but the thing that drives me most crazy is the politics here. So for him to get this done is a lesson for other people. **How do you find a common ground in a world that can't seem to do that?**

They're talking about building a pathway that would connect the college to the downtown. I think it will get more traffic when it's a nice, attractive feeling of safety. When the mayor took over, Caldwell and Canyon County had a reputation of being gang-infested. I grew up in Detroit, so that's a little out of touch with reality, but by Idaho standards this was a crime-ridden place and now it's actually a lower crime rate than Boise. When I tell people that they're shocked.

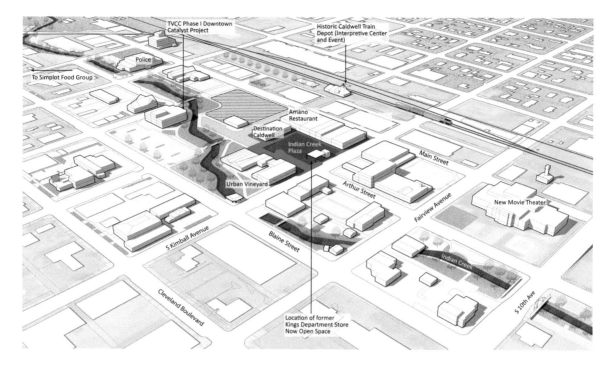

Steve Fultz, Caldwell Director of Economic Development

It's been a challenging ride from an economic development standpoint, because when we finally daylighted the creek and had the ribbon cutting, boom, we were already in a recession. The timing was horrible. We had huge vacancy issues in downtown and no money coming in. So one of the things we did at Caldwell Economic Development Council was to try to build up the most basic-sector employer. We targeted food processing companies and light manufacturing. We brought in a Foreign Trade Zone—the only one in the state. And then we aggressively went after some of these industrial users with the idea that they were going to be in the urban renewal district or the TIF (Tax Increment Financing) district. The industrial park is part of the urban renewal district.

We were able to secure Fresca Mexican foods, and Capital Distributing, and suddenly these Californian companies started coming in. We started seeing private investment that we could put in the downtown area. One of the most interesting ones to me was to me was Southwark Metal Manufacturing, headquartered out of Philadelphia. They manufacture ductwork and wanted a western presence. When they first came in, the CEO said, "I want to see your downtown." Everything was still pretty much a wreck at the time. I said, "Well, can you come back around Christmas time? I'll give you a tour at night." Because it was way too embarrassing. That was one of the things that stirred us to say, "We can't just have manufacturing or just go downtown. If we're gonna be a full-service community, let's be a full-service community."

The people downtown who were desperate to see revitalization said I was spending way too much time expanding the industrial park; the developers of the industrial park would say I was spending way too much time on downtown. But I think if you did a full analysis, you would see just what that industrial park growth did for the redevelopment of downtown.

There are still a lot of people with negative feelings toward urban renewal. Downtown property owners say, "You're raising my taxes." No, the tax rate is not going up. It's the assessed valuation of the property that's going up, causing, of course, the taxes to go up as well. Or TIF and the other taxing entities saying, "No, we're not getting our fair share." Our argument continues to be that had it not been for the urban renewal/TIF district there would have been no incentive for these guys to come out and be part of it.

Small businesses old and new in downtown Caldwell. (Left) Indian Creek Plaza during some of the programmed events.
Credit: (left) © Eleven One Photography; (right) Authors.

A Conversation

Skip Oppenheimer (SO) and Jeremy Malone (JM),
Developers, Oppenheimer Development Corporation

SO: This was the kind of opportunity where we felt if we could be the first to do something after a lot of years it would stimulate other development. We had done that in Boise with One Capital Center as the first downtown redevelopment project. And we recently have done that over at Oneida Falls and in Eagle, Idaho. We like the idea of downtowns. It's the heart of the city. If we can be a catalyst project developer working in public–private partnerships, that's not just decent business but it also creates a bigger impact.

JM: They [Caldwell leaders] wanted to do a new City Hall because, as in many small cities, their departments were in eight different buildings in a twelve-block area. They wanted to get everybody under one roof. Like many cities, they had hired a lot of consultants to come in and provide reports, but then you get this overwhelming 200-page document that nobody even wants to read because it's so intimidating. So they shelve it. That was the Leland Report out of Portland, this great report of what Caldwell could be, but there was less of an implementation plan. We responded to the RFP to create that City Hall block and it evolved into a Framework Master Plan.

SO: We said, "We can do City Hall but if it isn't contextually situated, you're going to end up with just a building. But if you create an overall downtown urban masterplan it's going to create context and attract others in a more thoughtful way."

. . .

We started with the City Hall, then the Treasure Valley Community College. But then, in the middle of all that, FEMA put all of downtown Caldwell and Nampa in the floodway. It was actually a modeling error, and it delayed the project by one year. We just hung in there. We weren't going to walk away and we knew the city would wouldn't walk away.

After we finally got things straightened out with FEMA, it was nine or ten months to get it built because Treasure Valley Community College (TVCC) had to open in August. The city wanted to own this as a public building. They had a lot of specialized stuff but they also wanted it to be designed in a way that if, God forbid, something happened to Treasure Valley Community College they would be able to use it as a public building. We acquired the ground and took the development risk and cost. And then the Urban Renewal Agency acquired the building about eleven months after opening.

JM: Obviously, there were economic changes in 2007, 2008 and 2009 and Micron Technology, which is based out of here, let a lot of people go and gave them an educational stipend. It wasn't enough to cover Boise State tuition but it could cover a community college. TVCC at that time was in a really small facility. They needed more space. The mayor had the foresight to say, well, it's the wrong time to build a new City Hall here but maybe there is a way that we can still get this building built in downtown and have them as a tenant. So, TVCC is a tenant of the building we owned.

Martin Hahle, President CSHQA (Architect/Planners)

When I first walked downtown, they were un-burying the creek, and digging and tearing up roadways and bridges and structures to make that happen. The city was talking to us about specific catalyst projects that they could do that would help energize development. They had a few key assets that they could play off of—their little train station building, which is quite unique, and some historic downtown buildings. It was an old, very traditional small-town downtown. They had City Hall and the police station, but those are spread out into multiple buildings. So they were trying to come up with ideas of what could we do besides opening up the creek. Because you'll have this creek, but there's not much there, you know?

We designed the City Hall, and we did some public presentations to the community to see if there was public backing. A lot of people liked what it looked like but I think they struggled to really get support to go spend a bunch more public money. They wanted something to make things happen down here, and they quickly saw that the City Hall wasn't the right way to go and that they were going have to approach this from a different angle.

At the time they had some kind of Mom-and-Pop sort of businesses that were down there, little shops that really don't generate a lot of revenue and traffic but that are nice to have. You want to add to them and build more. Ultimately, they found Treasure Valley Community College (TVCC), who was interested in potentially doing something. It was this public–private developer-funded effort to build the project and lease out what ended up being a three-story building. I was the architect of record. TVCC has a lot of classes there, and I think they generate quite a bit of traffic, particularly with evening classes. But in the end they didn't get quite the kind of traffic they had hoped for. A lot of the students take their classes and then leave.

The new Treasure Valley Community College building (visible at the bend of the creek in the center of the frame) brings an accessible new institutional anchor to downtown Caldwell along the new walking trail and restored creek corridor. Credit: Destination Caldwell.

Notes

1. According to the U.S. Census Bureau, Caldwell's population grew from 46,237 in 2010 to 63,629 in 2021. See https://www.census.gov/quickfacts/caldwellcityidaho.
2. Daylighting is a subset of the field of stream restoration and refers to projects that deliberately expose some or all of the flow of a previously covered river, creek or stormwater drainage channel.
3. According to the USDA and U.S. Census, Idaho has 1.4 cows per person.
4. Leland Consulting Group (2006: 12).
5. According to American Rivers, fewer than two dozen towns or cities have implemented an urban daylighting project in their downtown. For more information, see: www.americanrivers.org/daylightingreport.
6. 2020 data derived from the DataUSA website, accessed February 2023: https://datausa.io/profile/geo/caldwell-id?race-income-income_geo=incomeRace0&sexAgeRacePoverty=raceOption.
7. U.S. Census stories, accessed February 2023: https://www.census.gov/library/stories/2021/08/improved-race-ethnicity-measures-reveal-united-states-population-much-more-multiracial.html.
8. From the times of fur trappers to early miners and gold seekers, Idaho has always hosted extraction economies. "Pioneer" towns, in particular, were witness to severe exploitations and instabilities as a result of years of bonanza followed by rapid decline (Sparling, 1974).
9. LEED stands for the U.S. Green Building Council Leadership in Energy and Environmental Design rating system.

References

Alamilla, B. and Alamilla, S. (2020, January). D. Gamble, interviewer.

Brenner, N. (2014). *Implosions/Explosions: Towards a Study of Planetary Urbanization*. Berlin: Jovis.

Hoobyar, P. (2002). *Daylighting and Restoring Streams in Rural Community City Centers: Case Studies*. Seattle: National Park Service.

Leland Consulting Group (2006). *City of Caldwell Market Reconnaissance Report*. Portland, OR: Leland Consulting Group.

Nancolas, G. (2020, January). D. Gamble, interviewer.

Oppenheimer, S. (2020, January). D. Gamble, interviewer.

Pinkham, R. (2000). *Daylighting, New Life for Buried Streams*. Snowmass: Rocky Mountain Institute.

Race, Diversity, and Ethnicity in Caldwell, ID (2024). Retrieved from https://bestneighborhood.org/race-in-caldwell-id/

Sparling, W. (1974). *Southern Idaho Ghost Towns*. Caldwell: Caxton Press.

United States Census Bureau (2023). Quick Facts: Caldwell Idaho. Retrieved from www.census.gov/quickfacts/fact/table/caldwellcityidaho#

World Population Review (2024). Caldwell, Idaho Population 2024. Retrieved from https://worldpopulationreview.com/us-cities/caldwell-id-population

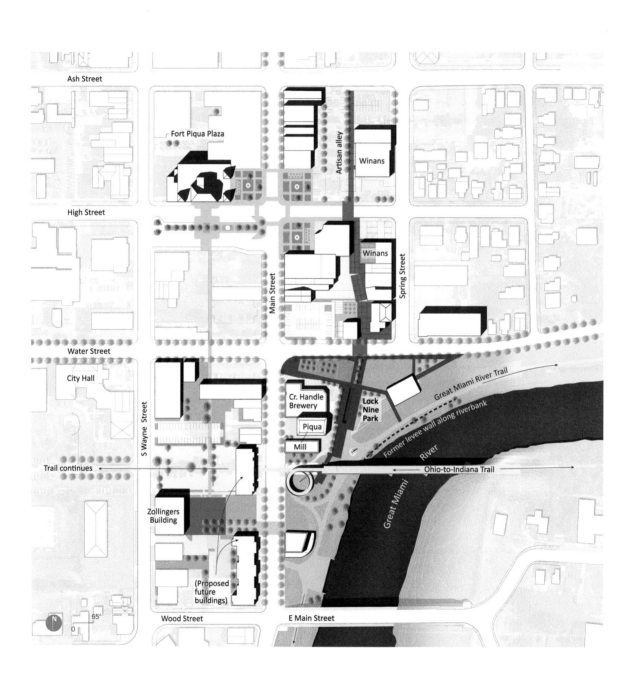

2.2
Piqua, OH

In the early to mid-19th century, an extensive web of canals, rivers, roads and rails linked the resource-rich heartland of the United States with the coasts. Over the course of just a few decades, entire settlements sprung up along these emerging networks that determined, to a large degree, winners and losers in economic development. Some communities, like Piqua in Southwest Ohio, were located at the intersection of several of these burgeoning transportation networks. Once famous for its annual "underpants festival" as a way to celebrate its heritage of fabric mills that made cloth for underwear and other clothing,[1] Piqua and its industries were enabled by the town's strategic location at the crossroads of canal and rail transportation—transport systems built up at the same time Dayton-born Orville and Wilbur Wright solidified the state's reputation as the "birthplace" of yet another modality—aviation.[2] Today, Piqua's population is just over 20,000 people (2024). The municipality is a northern satellite of Dayton, Ohio, and exists within the Dayton Metropolitan Statistical Area (MSA) where it remains at a crossroads, with interstates 70 and 75. The region continues to be driven by manufacturing, distribution and logistics in search of enterprise.

An early 20th century view of Piqua, Ohio that captures six transportation networks that were part of its early growth by river, canal, electric rail, steam engine, interstate and airplane.
Credit: Piqua Public Library/Piqua Historical Museum.

Historically, two primary canals laced north and south through Ohio. The Miami and Erie Canal ran along the western part of the state and the Ohio and Erie traversed the middle and eastern part. Construction of the 274 mile (441 km) long Miami and Erie Canal connected Toledo, Ohio to Cincinnati. Its construction began in 1825 and was more or less complete by 1845, passing through Piqua near the highpoint in the canal's elevation. The canal took advantage of water power from the Great Miami River and it provided the ability to easily transfer goods (and passengers) onto barges. Mills emerged along the canals and bars and hotels were frequently built adjacent to the locks because boats would typically take some time to pass through.

Unfortunately, the slow-moving canal boats (the average speed was a mere four miles/hour) were eclipsed by the railroad as a viable means of transport, and by the end of the 19th century, most canal waterways throughout the United States were abandoned. The "ditch that brought the world to the wilderness" became more of a liability than an asset (Gieck, 2007). By the time a massive flood devastated southwestern Ohio in 1913 there were few remnants of canal infrastructure remaining. Most communities had demolished their canals. In Piqua, they simply buried it intact.

The concealing of the canal erased an important historic artifact of Piqua's past. The only remaining evidence of the Miami and Erie Canal through Piqua were a few limestone fragments stacked adjacent to the Great Miami River inscribed with the text "Lock Nine." A once bustling railroad line and trestle that ran perpendicular to the canal also met a similar fate when that means of transport through downtown was itself eclipsed, this time by highways and semi tractor-trailers in the mid-20th century.

Piqua's center was isolated from the riverfront by another piece of infrastructure built shortly after the canal was buried. Following the massive 1913 flood, the Miami Conservancy District (MCD) built a steep revetment wall that followed the course of the river. As with many communities in America, this

View of the future site of Lock Nine Park in Piqua, Ohio from the elevated train trestle (now a regional trail) overlooking the Great Miami River and existing levee embankment.

levee served a critical function of protecting the town against flooding, but it also formed a sizable barrier that obscured the view of the river and detached the community from the recreational trail that lined its edge. Today the town is unearthing the lock and canal and re-grading this levee to reconnect people to the river in a new center of public life.

Piqua's riverfront project highlights what smaller municipalities are doing to celebrate aspects of their forgotten infrastructures to forge a stronger sense of place and enhance mobility. If by flight, canal, rail, trail or roadway, the city's transportation networks had always provided the platforms for private investment. However, by the start of the 21st century, Piqua faced economic decline and private-sector development lagged behind public-sector expenditures.

Past Futures

Piqua's leadership recognized that its downtown infrastructure and buildings had languished when compared to other communities its size. Although some municipal spending was directed toward beautification efforts like new brick sidewalks, historic lighting and continuous street tree planting, a county-wide branding study in 2010 revealed the palpable lack of confidence and skepticism in the minds of residents about downtown's status. Gary Huff, City Manager at the time, described it in this way:

> The community was faced with a reality-check following the branding study and this is what they thought of their community. It wasn't good. It was almost like a punch in the stomach to see that this is what you are, and unless we do things to change it, it's only going to get worse.
>
> (Huff, 2018)

Piqua's population has never been above 22,000 people, and it has long been overshadowed by adjacent communities like nearby Tipp City and Troy that fared better due to local circumstances, political fortunes and their closer proximity to the urban center of Dayton. Piqua is a mere seven miles further north from Dayton. Increasingly, however, small towns offer a lower cost of living when compared to larger cosmopolitan centers. Particularly post-COVID-19 pandemic, there are additional opportunities for remote work that make living in towns more viable as places of employment. Such a sentiment was expressed by Melissa Leingang, a young banker and active member of the region's Emerging Professionals Network:

> The reason I moved to Piqua was because it was the cheapest rent, and I could live by myself and not have to have a roommate. I found an apartment downtown and I could walk to work. My consolation prize for living in Piqua was I could live downtown and walk everywhere. If they would put a McDonald's and a grocery store near me (then) I would never need a car.
>
> (Leingang, 2018)

It's the character of place and its layers of meaning that unites generations. People of all ages are looking for a concentration of amenities in close proximity to one another and within a safe, walkable environment that doesn't require a car. Piqua's investments on the river signal this priority to make a special place for people to easily access—and feel a part of.

Locking into a History

In an effort to recognize their own assets following the negative county survey, Piqua's leadership embarked on a series of planning initiatives. Its 2006 Comprehensive Plan updated what was written in the 1970s.[3] A 2013 Downtown Riverfront District Development Plan followed that focused attention on a small section of downtown that had the greatest opportunity for transformation. And that was further advanced in a 2015 Downtown Placemaking Initiative. The strategies identified physical barriers to redevelopment, including key properties which needed to be remediated or redeveloped, and highlighted specific characteristics that distinguished downtown Piqua from other communities its size. Piqua's efforts concentrated reinvestment in a four-block area by the river that captured the town's historical legacy of the canal. Contemporary design could engage the past with new public spaces. The plans advanced an urban design strategy intended to spark private investment along its edges and beautify and sustain the downtown's open space network.

Residents of all ages gather by the limestone walls of historic Lock 9, now part of a series of new public spaces on the Great Miami River. Credit: Christian Phillips Photography/Gamble Associates.

The Placemaking Initiative was led by Cambridge-based Gamble Associates with CityVisions Associates and Booker Design Collaborative. Their signature design intervention was a reconfigured open space called Lock Nine Park after the former Lock Nine on the Miami and Erie Canal that it uncovered as its centerpiece. The 3.5-acre waterfront park serves as a trailhead to a 330 mile (531 km) regional trail network that runs along the Great Miami River. A second footpath, the Ohio to Indiana Trail, hovers above the riverside trail on the former railroad line that runs perpendicular to the river. The park space regrades the site's riverside contours and enables adjacent commercial activity right atop the levee (a condition rare along the entire length of the Great Miami River). Lock Nine Park incorporates the primary historical artifact of the lock by excavating it and removing significant sections of the steep, concrete-formed levee wall. The riverfront park reclaims Piqua's recreational edge as it stimulates the reuse of formerly industrial properties adjacent to the canal right-of-way.

Communities across the country are eager to showcase and celebrate aspects of their uniqueness. For Piqua, it was the latent transportation infrastructure that could make it distinctive. Unlike other towns that had fully removed their canals and dams—or elected to erect ersatz historical markers in their place—Piqua had covered its 19th century infrastructure without changing it. This hasty decision at the beginning of the 20th century enabled its preservation and reinvention later as the cornerstone of an inviting contemporary public space at the river's edge. The park's ribbon cutting took place in July 2024 and now serves as Piqua's largest gathering space downtown.

Piqua's Lock Nine and surrounding urban fabric circa 1900.
Credit: Piqua Public Library/Piqua Historical Museum.

The corridor's revitalization engages structures along the former canal alignment, so what were once the backs of buildings are being incrementally reoriented to become funky front doors for new uses facing the new open space and river. The canal no longer transports goods or passengers, but the corridor is a passage for people of all mobilities; it forges a contemporary connection between downtown and the Great Miami River Trail. As a publicly owned right-of-way, the old infrastructure corridor also accommodates new, forward-thinking infrastructure: the town removed overhead utilities and bundled and aligned them with new cable and high-speed internet lines resting at the bottom of the lock surface 14 feet (4.3 m) below the elevation of the park.

The Canal Traces a Return to the River

The memory of the 1913 Great Ohio flood lingers in the collective consciousness of the community, and many localities still suffer the detrimental effects associated with climate change and increased storm events. The presence of the levee required a design solution for Lock Nine Park that balanced protection from flooding with the promotion and extension of the trail path along the river. Ground penetrating radar initially identified the location of the buried lock walls and soil probes revealed that the historic limestone walls were in sound condition. The levee walls provided the conduit for a pathway between the downtown and the river's edge via a gently sloping ramp sandwiched between them. The removal of an eight-foot (2.4 m) high brick wall from the granite tombstone company that separated the park from downtown opened the space up to the river and an interactive water feature recalls the utility of the original spillway. A series of terraced steps negotiate the grade change in addition to the ramp. Interestingly, removing the steep concrete levee wall did not expose a flood danger to downtown. Tapering the topography actually increased the retention area at the bend of the river in the event of rising waters and the top of bank elevation is maintained.

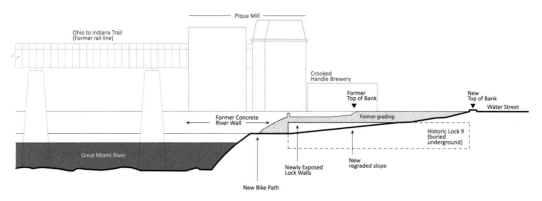

A cross-section of the former levee embankment and its new sloped/stepped-back profile that exposes the historic lock and enables passage during non-flood events.
Credit: Gamble Associates.

Buried for a century, the 19th century lock (seen in far center of frame) now organizes different kinds of public spaces and a new town common that overlook or step down to the river.
Credit: Christian Phillips Photography/Gamble Associates.

Ohio boasts over 800 river miles (1,287 km) on 14 watercourses throughout the state, and the Dayton region has over 330 miles (531 km) of continuous, paved trails along the river. The design of Lock Nine Park makes it easier for people walking, biking or running to negotiate the change in elevation and the terraced steps and ramp to the lower bike path along the river provide opportunities to enjoy the natural habitat and aquatic life. Plans are also underway to naturalize the riverbed. Lock Nine Park includes a performance pavilion, open lawn, overlook and tree groves that address heat island effects and anchor the open spaces. When large-scale events are planned for the performance area, controlled access points and tents will line the park's perimeter. Underground infrastructure enables lighting, sound systems and hook-ups for food vendors.

The Business of Small Town Momentum

One small, successful project can seed many more. However, regardless if it is new construction, historic preservation or adaptive reuse, there are risks involved. In small towns, where economic development has been dormant for some time, it can be a challenge for developers to take a chance with rebuilding and convince lenders to invest. Moreover, while there are many private-sector actors that will follow redevelopment trends, there are far fewer willing to lead the charge, especially when the market is untested and there is a scarcity of comparable projects with a viable rate of return. From the public sector perspective, years of defunding at federal and state levels has led to competing interests that often stymie investment in new parks when more pressing priorities emerge.

Still, there are advantages of investing in smaller municipalities. In a larger urban center, land costs tend to be greater, and the teams needed to reposition or redevelop properties are larger and more complex. Sometimes poorer neighborhoods near booming areas in large cities are exploited for access to cheap land (Heyda, 2024). In smaller cities and towns, land is still relatively less expensive. This lowers the barrier to entry and makes it possible for smaller developers to acquire and hold properties while financing is assembled and a viable development strategy emerges. There is less competition.

In smaller municipalities, even a modest-scaled building, or open space or infrastructure initiative can marshal a community's sense of pride and how other communities view them. A Dayton-based brewpub has recently occupied the site of the former granite tombstone business overlooking the river directly adjacent to the park. They did so under the expectation that the public investment initiative would be completed. And numerous historic properties along the former canal right-of-way are being adaptively reused. At the same time, progress has been incremental and residents are anxious to see more change more quickly. Some local developers seek added assurances from the municipality about public realm investments and streamlined regulatory reviews (the city recently developed a form-based code).[4] Others just desire less *bureaucracy*.

The 1913 S. Zollingers building in downtown Piqua. Located one block from the river, the 40,000 square foot former warehouse has been vacant for a generation and awaits new life.
Credit: Christian Phillips Photography/Gamble Associates.

Small Town Strategies

The American economic development model mandates that strategic public investment in a municipality's public realm stimulate private development around it, yet evidence of underutilized properties in Piqua's core remains. However, the downtown's restoration is bringing increased confidence and a number of modestly scaled structures are slowly being brought back into use. The buildings and the stories they tell are seen as building blocks with significant character for economic development (Alberts, 2013). By focusing attention and policies in a small area downtown and along the riverfront, Piqua is directing resources toward a longer-term vision. **Investment in open space is seen as a measure to resuscitate the vacant downtown building stock and generate new economic and design momentum**, a priority if Piqua's nearby residential neighborhoods are to stabilize and thrive.

Since American town public sector capacities are fiscally constrained, evidence of public investments can be drawn out as capital improvement plans take time to implement. As is common elsewhere, the private sector takes measures to shore up business capacities in other ways during the lagtime. Piqua has an existing, engaged private sector community that stepped up to establish a pre-development fund to enable transformation, with community and business leaders raising money as a first priority action item following the Downtown Riverfront District Development Plan. Initiated by the Piqua Community Foundation, the "I-Piqua Fund" raised over $1,000,000 in less than six months for gap-financing associated with new projects. Two building conversions, the ZOLO Marketplace (a rehabilitation of the historic, 1913 wholesale grocery warehouse) and the Crooked Handle Brewery (a rehabilitation of a former tombstone business overlooking the park) benefited from low-interest loans made possible by the fund to assist with site acquisition and building improvements. These catalytic projects are forming a critical mass of activity that neatly aligns with the urban design plan's aspirations. In the gaps between, strategic properties have been acquired by the city to remediate environmental issues and reposition them to attract new investment, or to guard against redevelopment that doesn't sync with the overall downtown strategy.

These planning logics follow trends seen in communities of all sizes. In this regard, Piqua's size seems less consequential. But when it comes to implementing a long-term economic development plan, size matters. Urban design projects take many years to orchestrate, design and implement, but once pieces fall into place, the momentum generated can go farther in a smaller setting—especially in recasting people's perceptions, pride and engagement with the town. As progress on Lock Nine Park and the canal right-of-way continues, mindsets are changing. One of the earliest signs was a 40,000 square foot former newspaper building which was converted into the flagship location for a fourth-generation family-owned business called Winans Coffee & Chocolate, increasing traffic downtown. An expansion in 2023 renovated an adjacent auto repair store for their coffee roasting. Other historic buildings are being adaptively reused for athletic stores and recreational gear. A former Edison power facility across Water Street from Lock Nine Park has

been cleaned and repaired for a new (yet to be determined) use. Next to Lock Nine Park is one of the city's oldest structures. The Piqua Milling Company along Main Street and facing the former canal has been vacant for generations. It is slowly deteriorating and awaiting new life as its owner looks for further evidence of change.

The municipality has been methodical in implementing the Downtown Strategy and Placemaking Plans, but despite indications of change, some property owners remain skeptical, revealing the tension between planning and implementation. The city invests in streets, sidewalks, infrastructure and the public realm in the hopes that private sector investment follows. In the current U.S. political economy, these investments generate revenue through increased taxes that then add to coffers that provide key services to support quality of life for residents. But until projects actually move into construction and evidence of physical improvement is tangible, many property owners are reluctant to invest. Meanwhile, governments remain financially strained and

(Left and center) View of the alley-like former canal corridor leading north out of Lock Nine Park where Winans Coffee & Chocolate transformed an old newspaper building into a regional production facility and where other local businesses are following. (Right) The vacant but stately Piqua Milling Company building, poised for redevelopment. Credit: (center) Wilson Reiser/Winans; (left and right) Authors.

Lock Nine Park
Capital expenditures in large infrastructure projects require a variety of state, regional, and local funding sources. *Source: City of Piqua*

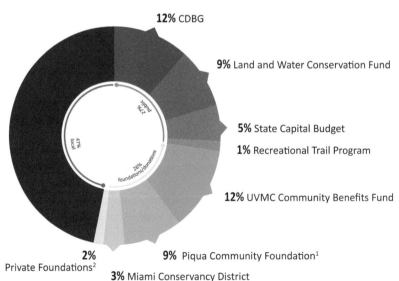

12% CDBG
9% Land and Water Conservation Fund
5% State Capital Budget
1% Recreational Trail Program
12% UVMC Community Benefits Fund
9% Piqua Community Foundation[1]
3% Miami Conservancy District
2% Private Foundations[2]

47% local
27% public
25% foundations/donations

[1] A Better Community Fund of The Piqua Community Foundation
[2] Lundgard Foundation ($40,000), Cargill Cares ($26,000), and a private donation

often don't even have the ability to uphold institutions and provide quality services for everyone. Timing is everything in the current context where towns with limited resources are forced to choose where to direct attention and investments. Unfortunately, no municipality—especially small towns—can really afford to disperse precious resources across expansive geographies. The alignment of public and private initiatives need to be highly coordinated to be impactful.

Financing for Lock Nine Park's completion ultimately required a multi-layered approach. Piqua established a Tax Increment Financing (TIF) district and attained funding from the Miami Conservancy District, various state Conservation Funds, two private foundations and a few corporate and individual donations. An integral component of their strategy includes extending investment further up the canal right-of-way and repurposing other vacant buildings along the canal corridor to complement new ground-up construction. The question that towns like Piqua are asking—even banking on—is whether repurposing their downtown's historic buildings into a mix of uses, including residential units, will sufficiently build on the existing businesses to anchor new kinds of community that attract and keep the next generation. Chris Schmiesing, Director of Economic Development and former city planner describes the small town rebuilding-precipice: *"It's about inertia, and we are at a tipping point."*

Perspectives

A Conversation

Gary Huff (GH), Piqua City Manager

Chris Schmiesing (CS), Piqua Economic Development Director, Former City Planner

GH: Obviously, we were really trying to reinvent the city because of deterioration that had occurred for years, and the question was how do we reinvigorate our nice central downtown area but also tie that into the river as one of our greatest assets. The studies and grants that we received helped us to determine that there are some really great things we can do not only for revitalizing the downtown but also getting people to live and work here and visitors to come here for all kinds of events. Now we're starting to see some projects happen, with the impetus of having this wonderful asset that really needed to be taken advantage of.

Somebody's got to step up and recognize the fact that change needs to happen, and that's really our role. I think the community finally got it that we've got to do something or we're just gonna die. I've seen it happen too many times. Towns and cities one day wake up and say, "What happened?" Well, you didn't do anything. You didn't improve your infrastructure, you didn't make changes to your community. With any city, without change, you die.

Our role has been to set the table. Our staff has done a good job of putting the information out and providing the vision and so far our elected body has been very supportive and provided all the funding and approvals for properties that we've purchased and made sure they got the type of development that's beneficial to the community.

CS: Part of the reason the i-Piqua model was put forward was so that we could do this in a sustainable manner, as opposed to some of the other fundraising successes we've had in the community where it's a "one and done" kind of thing. With i-Piqua there's a financing mechanism for folks to be able to influence positive change in their community in an ongoing manner, instead of needing to re-boot every time. And folks are beginning to gravitate to it but it's still a little bit unclear to them—like, which project am I investing in? It's a change in concept, so it's taken a bit more salesmanship to help folks wrap their mind around it.

GH: We're currently working twenty-seven active development projects and there could be a lot of opportunities for people to participate in whatever projects they feel are beneficial. I think the i-Piqua Fund understands their mission. They just finished raising the funds so we'll be bringing projects forward for their consideration and keeping things in front of them. We're talking about so many opportunities right now that you imagine if all of this comes to fruition it will really change the community in a positive way.

CS: The difference now, and this isn't to slight anyone in the past, is that the community is yearning to move forward, and there's leadership now at the city government level that's making it possible. Folks are empowered by that and they're seeing that the right things are happening, or that resources and projects or planning efforts are coming onto the table that they can get behind and push.

Janet Bly, Executive Director, Miami Conservancy District (MCD)

The Conservancy was formed after several floods in the 1800s and a massive flood in 1913. And while we're primarily a flood control organization, from the very beginning the leaders of this region intended the flood protection infrastructure to also be available for public space. But even with the flood risk reduced, people moved away from the small towns. Highways were built and the water infrastructure went out to other areas and the small towns along the river were not the most vibrant, attractive spaces to live in for a long time. The dirty industries stayed there and dumped their waste into the river, which made riverside towns even less attractive.

But things changed and people have again become attracted to the communities that are along the rivers. That means that all three components of our organization's mission are currently very important: flood protection, water stewardship and outdoor recreation. We still have to reduce the risk of flooding, because people aren't going to want to live in a river town that floods. They also don't want to be living next to polluted water, so our mission of trying to help promote the water quality is critical. And we want the river to be a bonus, not just a threat, because it offers outdoor recreation and community gathering spaces—it's really part of the identity of the community. Piqua, like a lot of communities in our corner of Ohio, has a really interesting

Recurring floods, like this one in 1913, made the Piqua region reconsider its relationship to the Great Miami River. Credit: Piqua Public Library/Piqua Historical Museum.

heritage and the river is in the middle of it, with the flood, but also the canals and the mills and the bridges. There's just a whole lot to the story of Piqua and the river.

When our system was originally built, there were levees to help prevent the river from overflowing its banks. And the amount of real estate that was available for them was quite limited because there were businesses right up to the water's edge. In some places, including Piqua, we have pretty tall, steep levees. They're like a mountain between the homes and businesses and the river. Over time, some of that real estate has opened up for us and we can set that levee back a little bit more. The height still has to remain the same to prevent the water from overtopping it, but if we can pull it back away from the river it actually improves the capacity within the channel itself and makes it more welcoming for the community to really engage with the river instead of having that steep wall. We're working on it little by little as opportunities arise.

Piqua's Lock Nine Park is different because there is a little bit of room to push the levee back and reduce the steepness. Reducing the slope of the levee invites people to sit and play on it, and it's less a barrier and more of a *slanted park*. This public space along the rivers that was part of the flood protection system became a great place to build trails, and during our stay-at-home period from the COVID-19 pandemic, the usage of the trails in Ohio has gone up 300 percent. People are able to get out and exercise without being in a crowd.

. . .

The Corps of Engineers builds systems very similar to ours, only they do it with federal money. But along with federal money comes, you know, the federal bureaucracy. We use their technical guidance and build to at least the same standards they do, and sometimes higher ones. But we are able to be a little more creative and get things done more quickly. The downside is that we don't have access to their money.

The regraded embankment and wide cascading steps of Lock 9 Park in Piqua reconnect people with the river.
Credit: Christian Phillips Photography/Gamble Associates.

There are about twenty conservancy districts of various sizes in Ohio created by law, and one size absolutely does not fit all. We follow state law, but we address our own regional issues related to water and form our own plans to carry out our mission. The public ownership of the land for the flood protective system has meant that once the community had the drive, the real estate was there. That's a challenge for a lot of communities. If you have to secure privately owned land it becomes a lot harder to create a riverfront district. Many communities also built their roadways right along the water, which is also an obstacle.

We have a program that we're calling the Great Miami Riverway, and it's ninety-nine miles of river. The watershed itself is over 3,000 square miles and while the work that we do encompasses all of the watershed, our focus is really on the river towns. Our system includes ten cities that have all banded together and said, "Let's create a regional destination so that somebody can come visit one community after the next and really get a sense of place that you might not have if you just visited one town on its own."

Kristin Booker ASLA, Principal, Booker Design Collaborative, Landscape Architect

My firm, Booker Design Collaborative, is mostly landscape architecture and urban design, and we work across all scales and project types. I've been in Louisville, Kentucky for over twenty years, practicing in both the public and private sectors. My office was part of the Placemaking Initiative for Piqua that recommended moving forward with the park as a Phase 1 implementation project, as we saw Lock Nine Park as the piece that would help to catalyze redevelopment and bring the community together by embracing the Great Miami River. We spent a lot of time in Piqua, many public meetings. Given the levee and transportation issues, a big piece was trying to figure out how to leverage this space to the greatest extent possible as a public open space along the river.

A lot of cities grow up around rivers, for obvious historic reasons, but, you know, rivers flood. There was a massive flood that was the impetus for the creation of the MCD. And, as engineers do, they designed something to prevent that flooding from ever occurring again, and they brought in a levee. The levee was very well designed and maybe a little *over*-designed. This kind of barrier works against human nature's desire to be connected to water. For us, the really big task was how to figure out a way to bring back a physical connection to the river in a very visible way, while still working with all the requirements of the MCD.

I just remember how workable MCD was overall. In my experience, agencies like this get really uptight about anyone messing with their infrastructure. But the Miami Conservancy District had a very open way of hearing our ideas, and then being flexible about the ways we could solve the issue of creating a greater sense of connectivity. And although it's been difficult from a funding perspective, they've helped us identify funds available through their agency that might be available to further the construction effort of the redesign of the park.

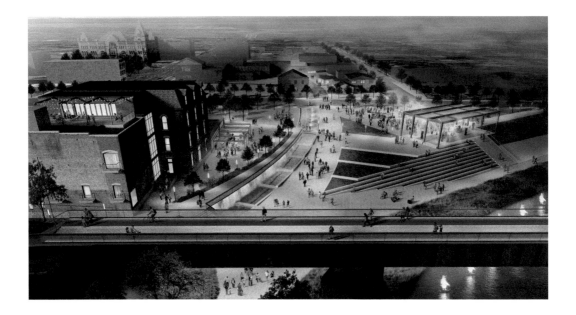

Aerial rendering of the initial design for Lock Nine Park in Piqua. Credit: Gamble Associates.

David Gamble was the thought leader in establishing the plan, and we thought a lot about the question of transportation. Like, the river used to be a way to transport goods, and then came the train, and then the roads. Now, the mode is the cyclist and the pedestrian. That's kind of new for this community. But we can tie that to the river, too, connecting people from Main Street and into the park and to the water. And then we looked at the adjacent buildings and their potential uses and how those buildings might put a front face toward the park and capitalize on that real estate. So that's how we came to determine the geometry, the paths, and the framework of the park's design.

The Great Miami makes this big turn to form a true oxbow that wraps itself around the heart of Piqua. And at that time when the main source of transported goods was on water, a really cool engineer decided to put a canal straight through the town rather than go all the way around the oxbow. So a shallow canal—like in Venice—came to exist in Piqua, Ohio. In our initial design for Lock Nine Park, we created a curvilinear new wall that reflected the notion of the canal wall, with the assumption that the original infrastructure was gone. But then, during a public planning board meeting, the town historian Jim Oda, said, "Why don't you just excavate the existing canal? It's still there." And we didn't know that; we assumed it was demolished! We had to bring in a ground-penetrating radar to actually identify that this 100' long (30.5m) by 14' (4.3m) wide x 14' (4.3 m) deep lock wall was still there. That was a dramatic moment. David and I had had productive conversations when the design was underway about reinterpreting the lock as this beautiful curved water feature that helped to connect people down to the edge of the river. Along the way, we both had a sense that was like "Man, should we not just put the lock walls back exactly where they were?" But who would have ever known that the lock walls would be intact beneath the earth and that all we had to do was dig them out to reveal them? That's pretty cool.

. . .

We need to feel a sense of connectedness to nature, but the design process tends to peel away nature versus holding it up and keeping it. It's one of the first things to be pushed aside. Some of the best urban designers and landscape architects working in public open spaces are the ones that can hold on to a true connected sense of nature. One of the areas of focus we had when developing the park design together was how to create a riparian edge that had a more naturalistic look and feel, so that when people were along the river they really felt like they were in a more naturalized setting. That edge condition also prevented erosion and created habitat for wildlife. But landscapes grow, and they evolve and change and they have to be managed. There's a considerable amount of maintenance that nature requires in these constructed settings. I think that's one reason why green infrastructure and ecology-minded design processes get halted or stunted. Our current mindset is often like, "What's the [environmental] return for our investment in caring for the landscape?" And if you don't value that return, you probably aren't going to invest.

Architects will say that the trees are getting in the way of everything, that they're just a shield covering their building. But people want to be in shade. We think of trees as buildings and about how they frame spaces, not detract from them. The amount of detailed landscape plantings in this park is really manageable. There are a lot of open lawn panels, plazas, and spaces and surfaces that don't require a great amount of care on a regular basis.

. . .

There are a couple great surprises about Ohio that I could easily compare and contrast with other states and towns that I've worked in. Everyone in Piqua was so excited and open to the ideas that were being presented. And I think they were really appreciative of the intentionality that the process provided. The levels of involvement and attendance at the community events were incredible. We did a lot of different kinds of engagement. The one that shocks me still is

Piqua residents and civic leaders during one of the tours of the town's vacant historic buildings.
Credit: Gamble Associates.

when Chris Schmiesing and David Gamble orchestrated a tour of a number of vacant buildings around the park that had adaptive reuse potential. And in that four- or five-building tour, there were like, seventy-five or eighty people. They couldn't wait to get back into these buildings because they hadn't been in there since it was business X or Y or Z. And everyone was feeling so nostalgic but optimistic at the same time. I think the community is craving something nice, and out of that will come a lot of pride and ownership. We're not just making a park—there's so much more to it than a pretty open greenspace.

A Conversation

Steve and **Sally Watson**, Owners of Tipp City Mill (in Tipp City, south of Piqua) and Piqua Milling Company Building (adjacent to Lock Nine Park)

Steve and Sally Watson have owned the Piqua Mill Building since 1991 when they paid $46,000 for it. It has been vacant since that time. They own and operate a number of successful enterprises in Tipp City, 12 miles (19.3 km) south of Piqua, including a mill that functions as an event space and a former hotel that is now an antique, clothing and art store.

Steve: I'm born and raised here in Tipp City. So, the community is near and dear to my heart. When we started, downtown was at risk because the big box stores were coming in and putting our hardware stores out of business, and it was just leaving a bunch of empty storefronts.

Sally: This was more than 35 years ago. We started buying property in 1987.

Steve: It was at risk. And it was just pulling on my heart strings, like, I need to do something about it and personally take some responsibility and start making something happen. It got to the point where that became a whole-hearted kind of thing. I just kind of put all my eggs in one basket, and that was downtown, downtown, downtown.

I took a single-handed approach to get things turned around down here. I became like a one-man chamber of commerce. I created all the events downtown. Early on that's all I did—I was like a fulltime downtown organizer/director. The events took all my time and money. I wanted to make a drum beat, a trumpet blast, and come up with neat activities that would attract people to come downtown.

I also did that to attract people from outside of the community. Doing it just for the people in the community wasn't going to be enough to stimulate the downtown like it needed. I had a broader marketing plan to attract farther out. But because it was a merchant-based initiative, the basis for all the events were the downtown shopkeepers. It wasn't somebody from outside organizing an event they didn't have a vested interest in. Everything was from the perspective of merchants' vested interests.

Sally: People come from all over to shop here; it's a destination. But also, now we have a lot more local people, and whenever they have company visiting, they all show off their downtown.

Steve: Once you become a destination the marketing isn't really needed. So the sacrificial years of spending a lot of money on advertising and promotion and marketing stuff are over now. We were trying to create a cultural marketplace where history and art were the handshake and to develop economies off of those two areas. I was antiques and Sally was art. We took a building down the street and put an antiques mall in and at the same time came over here to the hotel and put an art gallery in here. Artists are always looking for places to showcase their work, so if you have a long-range goal and you want to do it in phases they're an easy way to get things moving. And often developers take these large empty shopping malls and turn them into a flea market mall. That's how we jump-started downtown Tipp City. We branded it with antiques because the cause behind it all was historic preservation, and antiques fit that theme and reinforce it. My dad had a furniture store downtown here and he had a little brick building that he used for warehousing and I talked him into letting me use one side of it to open up an antique shop. But then I would go to New England and seeing these little villages would get me all excited, looking back to my hometown thinking, "Why can't this look like that?" I started seeing this as a Midwest-New England town. And that's where the inspiration came from.

My thinking all along was to strengthen the preservation ethic in the community. People weren't interested in history. Part of that goes back to our public school system where history and art aren't primary courses. So part of the thinking was to start showcasing the town's history, its heritage. I identified the canal as the town's birthright and felt that, if you're going to do this, you go back to the beginning. And so we branded all the marketing around old Tippecanoe and Tipp City as a canal town. The creation of the Canal Days event was going to be a way that we were going to educate people about the town's heritage. Now all of the elementary classes in the public school system come down here every year for an historic building tour of the downtown. It's a cultural strengthening of the community, preserving our history.

We're talking about reinventing community. *What is community?* It's close-knit. It's relationships. It's meaningful and intimate. But we've become a transient culture. People are scattered and spread out too far from one another. We have to call people back home. You have to adopt a close proximity lifestyle where you can draw a little circle around a small little area, like the downtown here, where you live and work and play. And both husband and wife have to be within that little circle. The spouse can't be working in another town doing something else. Sally and I, we live, work and play day in and day out. We've been going to work together for 30 years. That forms the seed for genuine community.

Sally: Six out of six of our children live nearby.

Steve: There's an old saying, "Never move so far away that you can't see the smoke from your parents' chimney."

And so all we did was basically reinvent community just like in the old days. It's a neighborhood. Early on down here when we still had the hardware store and the grocery store I never needed a car. I was always on a bicycle or walking. Everything was already down here. Now, every time I need even a screw I need to get in my car. But those are the kind of things that we need down here. If you don't have basic needs built into that area where you live, you're just offering luxury stuff versus selling something that somebody needs. That lowers the frequency that a customer needs to be in your face. **That's why a tourist economy can never be genuine, because nobody is getting to know anybody.**

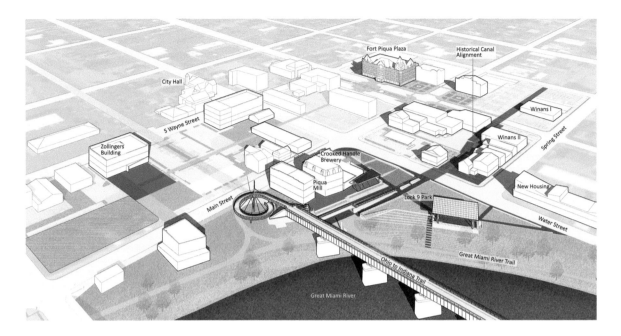

When you're developing an area you have to get those things built in as basic needs. To me, something like a micro-brewery is great but that's not basic, you know? That's entertainment and a luxury. It helps, but it's not an anchor.

. . .

Steve: That's why I don't worry about that Piqua Mill sitting up there. Everybody wants to know when its transformation is going to happen. Well, my thing is just sit tight, don't go up there and spend a lot of time and money. Wait for some collateral development, something outside me to happen to get the ball rolling. It's logical to wait and time it to when the city is doing their thing. Then I can make my connection at that point. It could well be, after all these years of waiting for a starting point, that the collateral development is the replacing of all of the (overhead) utilities. The city's doing something and it's causing me to do something, and you can look at all that energy that's created right at that one spot. And now, with the purchase and rehabilitation of the Zollinger's building across Main Street, you've got a situation that creates a harmonious energy.

The thing is, I didn't sit around and wait for some collateral development to happen here in Tipp City. It was the total opposite. But I know where I'm at and for me to go up there and make something happen, to not wait on the city and just start doing stuff, well, I know that's gonna take a lot of time and a strong commitment. Relying solely on my own initiative and resources to get something to happen here in Tipp City just about killed me. At forty-two, I was about dead and had to have open heart surgery. I don't have to do that there like I did here. I think that when the city starts relocating the utilities people will see something physically happening, and that will make all the difference.

Did you see that picture of that big three-story flag I have above the sofa? Occasionally, I hang it from the gutter of the building across the street. It's stunning. People just flock over there to have their picture taken in front of it. The other morning I thought about hanging that flag on the Piqua Mill building, just to spark something. I know it's gonna happen—at some appropriate time that flag will be on that building. It's the rebirth of Piqua!

Notes

1. Piqua and its underpants history is also the inspiration for Dav Pilkey's popular *Captain Underpants* children's book series.
2. In 2003, Congress officially named Ohio the "birthplace of aviation" in light of the fact that Dayton was the home of Orville and Wilbur Wright who are credited with designing and manufacturing the first flying aircraft. According to NASA, Ohio has twenty-three residents who have gone to the moon, more than any other state, and the Wright-Patterson Air Force Base, just south of Piqua, is the largest employer in the region, with over 30,000 military, civilian, and contractor employees.
3. Entitled "Plan it Piqua–Preserve the Past Empower the Future," the Comprehensive Plan Update recognized that Piqua's tight historic downtown core and expansive river frontage were two aspects of its identity that were being overshadowed by aging infrastructure, declining neighborhoods, and consumptive land patterns on the periphery.
4. Developed by Austin-based Code Studio, Piqua's City Commission passed *CodePIQUA* in April 2023 to streamline the development process and stimulate growth.

References

Alberts, B. (2013). *Riverfront District Development Strategy*. Piqua: CityVisions Associates.
Gieck, J. (Director) (2007). *Ohio's Canal Era: A Companion to A Photo Album of Ohio's Canal Era 1825–1913* [Motion Picture].
Heyda, P. (2024). *Radical Atlas of Ferguson, USA*. Cleveland: Belt Publishers.
Huff, G. (2018, October). D. Gamble, interviewer.
Leingang, M. (2018, August 16). D. Gamble, interviewer.

Section 3 **Metroposts**

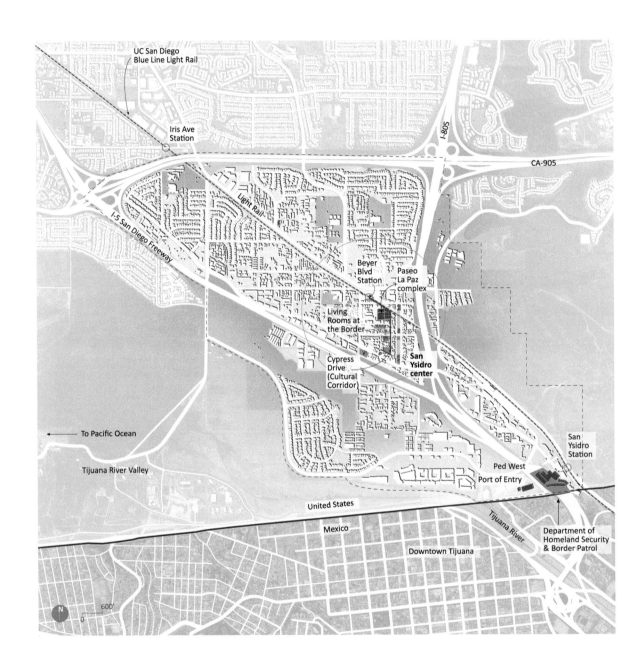

3.1
San Ysidro, CA

The *Atlas of Land Entry Ports on the US-Mexico Border* identifies fifty-two locations for legal land crossings, of which twenty-six are designated ports of entry (Rose and Davidson, 2010).[1] The largest of these is located 19 miles (30.5 km) south of downtown San Diego, California in the small town of San Ysidro (pop. 28,008) bordering Tijuana, Mexico. San Ysidro has had a land border inspection station since the 1870s, but by the 1970s the infrastructure was woefully inadequate to address the increased flows of people and commerce that a 21st century international border demanded. Beginning in 2009, the Federal Government completely modernized and expanded the international inspection facilities on both sides of this border and overhauled it with new administrative and operational facilities. Most of the traffic is northbound into the United States, and San Ysidro accommodates over 70,000 cars and 20,000 pedestrians *per day*. While the influx of people is a critical economic engine for the San Diego-Tijuana region, its impacts on the town itself are uneven. The final phase of this immense decade-long infrastructure construction project concluded in 2019 at a total cost of three-quarters of a billion dollars.

San Ysidro is defined by the Mesa Hills on the east and the Pacific Ocean on the west. But it is really the north–south highways leading to and from the port of entry which delineate the place. Interstates 5 and 805 bring access to one of the largest agricultural communities in the country, but they also hem the small town in, and San Ysidro bears the daily brunt of the adverse effects of the highways. Tens of thousands of idling cars increase pollution and reduce air quality, in addition to the other ecological crises a massive border wall produces. Moreover, despite the town's proximity to the affluent San Diego metropolitan region, San Ysidro is one of the poorest communities

San Ysidro Port of Entry funnels thirty-two northbound lanes of traffic from Mexico.
Credit: Photograph © Harry Connolly.

in California. The median income in the town is 60 percent less than the rest of San Diego County. Most census tracts in San Ysidro qualify as an "Area of Persistent Poverty" and an "Historically Disadvantaged Community" (County of San Diego, Health and Human Services Agency, 2022).

Negotiating Scales

For fifty years, and in the shadow of this vast physical infrastructure, an entrepreneurial non-profit in San Ysidro has been improving the region's *social* infrastructure. Casa Familiar enhances the quality of life for low-income individuals and families with wrap-around social services that combine advocacy, education, affordable housing, economic development and the arts. Their most recent built project, called "Living Rooms at the Border," challenges notions of traditional density and brings together new affordable housing construction, adaptive reuse, flexible communal and cultural spaces, and a multiuse complex in a dense neighborhood condition. Although the architecture is contemporary in its expression, the urban design of the campus is seamlessly woven into the fabric of the existing neighborhood: it fits. The project, together with many other initiatives by the non-profit, highlights the increasing role that Community Development Corporations (CDCs) have in building neighborhoods, via core provisions like housing, and by providing services that stabilize their host geography. **Living Rooms at the Border is a hyper-local, site-specific response in the face of a vast, politically charged international border**. It exemplifies a practice of "critical proximity," what architect Teddy Cruz and political theorist Fonna Forman refer to as a way for designers to operate in sites with intensive global political economic dynamics. They achieve this by zeroing in on the local spaces where those forces touch down (Forman and Cruz, 2024).

Technically, San Ysidro is not an autonomous town, but a jurisdiction of San Diego, California, annexed in 1957. But culturally and physically, the community of less than 30,000 people evoke a distinct place. It is included in

The local/global San Ysidro neighborhood/ border context. Credit: Photograph © Harry Connolly.

this volume of towns for that reason, and because of the intense geographic negotiations it balances. **San Ysidro is a small community on a global border**. Equally importantly, it is a marginalized district in the shadow of very large, more affluent city and metropolitan region—it's what we call a *metropost* town. The extremes of scale and stance create both obstacles and opportunities for creative design and local governance overlays. This case illuminates how reassertions of the *human scale* can inflect both design practice and urban form. The project of re-centering people in San Ysidro has mediated radically diverse footprints that are otherwise too big (the border crossing) or too sparse (a disinvested urban infill condition), with meaningful results.

As described by Roger Bruns in *Border Towns and Border Crossings*, the U.S.–Mexico border is "far more than a line or markers or fences that separate two countries. It is history and commerce and culture, as well as a minefield of political tension, fear, violence and exploitation … a landscape of desert and waterway that has been traversed for centuries and coveted by many" (Bruns, 2019). The dividing line between countries has been especially contentious in recent years, as immigrants seek refuge in the United States from the south and as trade continues. In light of the influxes, San Ysidro is the busiest border in the Western Hemisphere. It must process the 90,000 daily commuters crossing between Tijuana and San Diego, and migration numbers are projected to increase substantially in the future. In the past, individuals passing between countries had to endure hours in traffic waiting for U.S. Border and Customs officials to process them. A redesigned border crossing facility sought to expedite the process and humanize the experience.

Leading by Design

The redesigned facility was a bi-national effort between the United States and Mexican governments. The San Ysidro Land Port of Entry Expansion Project was a ten-year, three-phase effort involving the rerouting of highways and access roads, demolition of outdated facilities, expansion of inspection booths and administrative offices, increases in connectivity for pedestrians and general improvements to the experience for those who pass through. The project was part of the federal General Service Administration's (GSA's) Design Excellence Program established in 1994 to ensure high-quality design and construction projects by employing a rigorous pre-qualifying and peer review process. Firms must follow the agency's established design principles. Similar to the Walton Family Foundation Design Excellence Program in the Bentonville, Arkansas case in this book (but publicly led here), the initiative foregrounds the importance of design and the manner in which it can align with local community goals for specific projects. Seattle-based architects Miller Hull were selected to advance the immense infrastructure project, beating out larger firms with deeper portfolios for the prominent commission.

Around this time, the national non-profit housing developer Enterprise Community Partners, Inc. (Enterprise) was establishing their own design program called the Rose Architectural Fellowship. The program

The social services building of Casa Familiar nestled in the neighborhood in San Ysidro, CA. Credit: Photograph © Harry Connolly.

pairs young architects with community organizations around the country to provide non-profits with added capacity to bolster their community engagement efforts. Since 2000, every year a small number of designers are selected to work for two years in different locations around the United States. Non-profit partners apply to Enterprise with the hopes of landing a coveted, community-minded architect to assist them with their mission. San Ysidro's Casa Familiar was fortunate to be assigned one in 2001 (Swenson, 2020).

The Juarez, Mexico-born architect David Flores was selected by Enterprise to work with Casa Familiar for a role that would shape the port of entry undergoing redesign. Working as part of the Community Advisory Committee assembled to provide feedback through GSA's Good Neighbor and Art in Architecture programs, he advocated for higher design quality and a more humane experience for the high number of pedestrians that cross back and forth between the two countries. He was drawing on his own experiences growing up in a family divided across the border (Swenson, 2020). David used his design skills to shape the configuration of the infrastructure, including a second new pedestrian bridge called PedWest. In addition to his advocacy role on the committee, he analyzed the environmental factors that were impacting the small community and helped design affordable housing for the non-profit. Following the conclusion of his two-year Rose Fellowship, David remained on the staff of Casa Familiar for many years, becoming an integral member of the team and, eventually the Director of Development. His role demonstrates the importance and value of lived experience as a crucial dimension of professional "expertise."

The Community Builders

CDCs are created to support and revitalize communities, especially those that are impoverished or struggling. Casa Familiar is one such entity and although much of their funding comes from local governments and state and federal grants, CDCs are 501(c)(3) non-profit tax-exempt organizations. They may receive donations and grants from both public and private sources. Beyond building or rehabilitating housing provisions, many CDCs have expanded services to do much more. They fill the void of social services that municipalities cannot undertake, as privatization and the general erosion of public-sector capacity has rendered many municipalities unable to address the fundamental issues of housing insecurity and poverty.

Some CDCs get their start by developing housing first and then they move on to meet other needs such as day care, job training, health care or English as a second language programs. Others begin with these services and then move into the housing sector which relies less on grants for operations and administration (Stockard, 2023). A strong housing portfolio can provide stabilized cash flow which funds the non-profit to do more things. This is the path that Casa took under the leadership of its former Executive Director Andrea Skorepa. Since their establishment in 1973, Casa Familiar has grown from serving only Spanish-speaking clients in San Ysidro to providing for all South San Diego residents. Their comprehensive services align neatly in their mission statement: "To enhance quality of life for low-income individuals and families through advocacy, social services, education, affordable housing, arts and culture and community and economic development" (Casa Familiar, 2023).

Asserting the Local

The San Ysidro neighborhood where Casa Familiar is situated is dense, with closely spaced homes on small blocks less than 160 feet (49 m) wide. An eclectic commercial corridor along West San Ysidro Boulevard parallels Interstate 5 where it meets Interstate 805 and the southern edge of the neighborhood. Casa Familiar recognized that arts were essential to the cultural life of the community, and that the existing fabric of the town held local histories of place. One of their projects was an adaptive reuse of a 1929 dry goods store along West San Ysidro Boulevard that they turned into a concert, performance and art gallery. Part community center, part design venue, "The Front" is not static; the flexible space holds workshops, lectures, book readings and podcasts centered on bi-national life. There are apartments on the second floor.

The Front, San Ysidro's arts and culture venue run by Casa Familiar. It occupies a restored 1923 dry goods building and includes two housing units on the second floor. Credit: Photograph © Harry Connolly.

Casa Familiar expanded their development arm in tandem with their other services. They partnered with larger developers to build affordable housing at scale, as housing shortages in San Diego mounted pressures for more development around transit stations. These larger projects provide a more stable income stream for the CDC that in turn subsidizes other endeavors. To guard against gentrification, the organization purchased another key property in the neighborhood under a Community Land Trust (CLT). The property holds potential to add more affordable housing to their portfolio and ensure control well into the future.

Casa Familiar's earlier work involved small-scale infill projects that challenged existing zoning codes by paradoxically drawing from what had been allowed in prior eras—and what had been valued most: the town's small connective alleys and corridors, called *paseos*. Casa's long-term urban design plan envisions the development of a "cultural corridor" to link the organization's projects together. The corridor is designed to extend from The Front along San Ysidro Boulevard to San Ysidro's light rail station approximately one-third of a mile (0.5 km) away. Although the space is little more than a concrete alleyway today, Cypress Drive will be a pedestrian-oriented, resilient landscape inspired by the many verdant and social *paseos* found in the area. This future arts corridor is parallel to and one block removed from San Ysidro Community Park, which has the town's library, senior center, recreation center, civic center and playgrounds. Together, the framework secures a system of urban insertions that remain incremental while maintaining the scale and grain of the district. The right-of-way stitches together the long-term efforts of the CDC with each other.

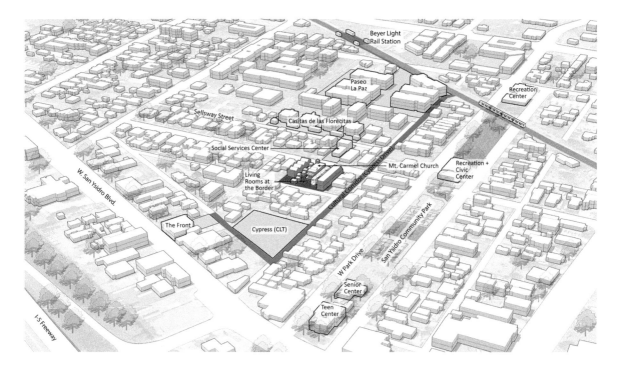

There is one more local dimension of this incremental approach to rebuilding, where *design engagements* have also been manifested over time between the San Diego social design practice Estudio Teddy Cruz+Fonna Forman with Casa Familiar and residents of San Ysidro. A framework put forward by this team reimagined the political processes in place in San Diego. Cruz and Forman leveraged the capacity of their academic appointments to expand partnerships and challenge regulatory structures that were more focused on economic development than on engendering socially rich spaces.

Contemporary Contextualism

The renowned team of architect Teddy Cruz and Fonna Forman, a Professor of Political Science and Director of the Center on Global Justice at the University of California in San Diego (UCSD) began collaborating with Casa Familiar to develop a pilot project for San Ysidro in 2001. Together, the two have spent decades working closely with Casa Familiar and San Ysidro residents to address abandoned and underutilized land in the neighborhood, and to confront some of the assumptions hindering creative new models of urbanism. Like Forman, Cruz is an educator and serves as Professor in Architecture and Urbanism in the Visual Arts Department at UCSD. Their research-based political-architectural practice complements their teaching and writing, affording them a critical platform to connect the UC institution with the surrounding community. One aspect of this platform is the design lab Cruz and Forman run that bridges the university's system of community engagement, called "Community Stations," and the Center for Global Justice that Forman directs. Cruz and Forman argue that *meaningful social justice lies*

3.1 San Ysidro, CA

in the ability to localize global dynamics (Forman and Cruz, 2021). Their design lab bridges broad scales to connect "top-town resources to the bottom-up intelligence of communities"—with the idea that the community itself could lead in the classroom to elevate research with the kinds of deep place-based socio-political knowledge that only residents and activists working on the ground will know and have experienced.

Unlike a more traditional community design center that functions as a service outreach arm for the university, Cruz describes their university-affiliated work this way, "It's not the university planting their flag. But we become a mediating agency between the community and the institutions. The center is a research-based practice that is inside the university so that enables us to do a series of triangulations that otherwise would be difficult" (Forman and Cruz, 2021).

Estudio Cruz+Forman's largest built project with Casa Familiar is "Living Rooms at the Border." The project arranges a collection of buildings and programs in an open yet compact pattern that challenges traditional notions of density. Living Rooms adaptively reuses an iconic church and arrays community-based programs around it, drawing together the adjacent buildings.

Located directly across the street from Casa Familiar's main administration building is the 1927 church that was rehabilitated as a community venue. A narrow two-story building adjacent to it houses affordable apartments for families, and the other side of the former church contains smaller rectangular pavilions for social service programs. The simple volumetric buildings are pared back in their forms and the ensemble is organized by narrow walkways and gardens, recalling the spatial characteristics of the existing neighborhood *paseos*. It's the spaces *between the buildings* that make the project particularly contextual. The urban design project was nearly two decades in the making. The ideas were first exhibited at New York's Museum of Modern Art in 2001. The built version opened just as the COVID-19 pandemic took hold in 2020.

Living Rooms at the Border is an act of urban acupuncture. Delicately inserted community-based programs fit seamlessly into the tightly knit urban neighborhood. The project effectively reinforces the neighborhood character already there, multiplying its impact. Despite its intimate contextuality, the designers had to push back on local codes that were paradoxically set up to encourage *bulkier forms*—for higher density and larger returns—rather than

The bold colors and *paseo*-like qualities of the buildings and open spaces at Casa Familiar's Living Rooms at the Border designed by Estudio Teddy Cruz+Fonna Forman.
Credit: (left) Casa Familiar; (right) Authors.

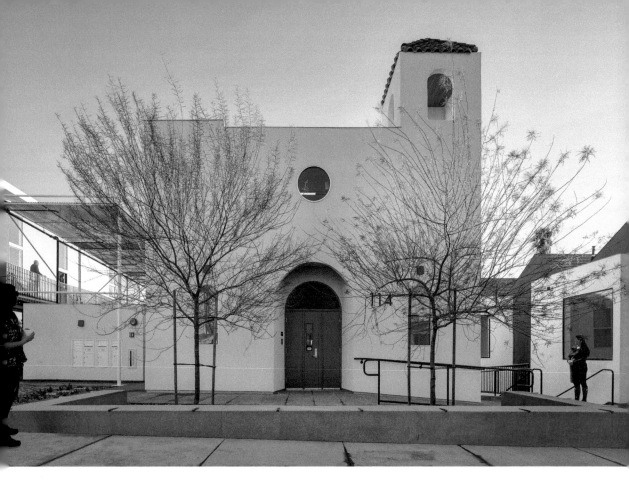

The renovated Mt. Carmel Catholic Church in San Ysidro, California, part of "Living Rooms at the Border," a collection of projects with Casa Familiar designed by Estudio Teddy Cruz+Fonna Forman. Credit: Estudio Teddy Cruz+Fonna Forman.

being set up to support appropriate local ways of making and inhabiting space. Like the CDC itself, the project combines many things together with affordable housing, studios, a performance pavilion and social services. The physical forms are plutonic and unabashedly contemporary with bright colors. Rather than attempting to look historic, the small village becomes a confident extension of the neighborhood rather than a threat to it.

The Living Rooms project was one part of a series of proposals called "micro-policies" put forward by the designers to assert "urbanisms at the scale of the parcel." Other ideas defined a "distributed land trust" that would help individual owners develop accessory dwelling units (ADUs)[2] and other improvements on their property. It also rethought tax-credit systems that promoted development of large buildings of fifty units or more, something that was not allowed in the code. As an alternative, the team proposed that those fifty units could conceptually be redistributed across fifty much smaller parcels, as accessory dwelling units to fit the existing fabric (Cruz and Forman, 2022).

Transit Orientations

California is notorious for its congested, car-driving culture. But the state has been augmenting auto-dependent infrastructures with investment in light rail and other public transit systems.[3] There are four rail lines that extend outward

View of Paseo la Paz, 137 affordable housing units completed in 2020 directly adjacent to the light rail station, in coordination with Casa Familiar. The 26.3-mile (42.3 km) UC San Diego Blue Line extension of the Metropolitan Transit System (MTS) opened in 2021 and connects the Port of Entry in San Ysidro with San Diego to the north.

from the urban center of San Diego (pop. 1.38 million) and it's the Blue Line that links southward to San Ysidro and the port of entry. The enhanced mobility brings access to the nation's front door with Mexico, but it also brings development pressures, intensifying the housing crisis in southern San Diego and areas nearby. New transit-oriented developments are rapidly emerging around the stations at a scale more akin to downtown San Diego than to the one- and two-story fine-grained fabric of San Ysidro. While larger housing projects mean that Casa Familiar can partner with more mature development companies, it also brings challenges to how affordability is calculated, and for whom. The CDC is creating strategies that ensure residents can stay in the neighborhood as land prices escalate due to the integration of transit and growth. Their CLT is one of these strategies, and the CDC's direct partnerships with larger developers have been another way to ensure a seat at the decision-making table to control where and how higher intensity design happens.

Perspectives

Katie Swenson, Enterprise Rose Fellow, Former Vice President of Design & Sustainability at Enterprise Community Partners; Senior Principal at MASS Design Group

I came in 2007 to run the Rose Fellowship Program. We looked for partners with an appetite and ambition around a strategy that had to do *specifically with design*. Most of these organizations are social service groups. In their real estate departments, we would ask "How are you thinking that having a designer on your staff would really amplify your mission?"

Casa Familiar and San Ysidro were early in the program. Their organization is now fifty years old. Their director, Andrea Skorepa, was a powerhouse. She was a dedicated, fierce advocate for her community, very creative and hardworking. She had caught the design bug from Teddy Cruz, and they shared a love of ideas. They had been scheming on Living Rooms at the Border. It just opened in February 2020, like twenty years later. That project got shown at the Museum of Modern Art ten years before it was built!

David [Flores] had started a project through Casa, which was doing research with the University of San Diego and the University of Washington about air quality. The cars [at the border] get backed up; they just sit and idle in this little town, and the town is in a valley, so it gets covered [in particulate matter].

This is David's life's work. He was born in Juarez, Mexico. He came to the U.S. with his parents and his aunt when he was nine years old, and the family was living on both sides of the border and he felt wrenched apart. When I was there, we heard from many people about family separation. The trauma that that word glosses over is just so extreme. David and his wife Lisa were looking at the mission statement of Casa Familiar, which was something very generic, something like "Casa Familiar fights for quality of life for the residents of San Ysidro." They were talking about changing it to "Casa Familiar fights abandonment." The idea of abandonment was a central theme.

All these families are all tangled up on every side of the border. David says "Love doesn't know borders." I'm so drawn to David's story. His father can't come to the U.S., but they can go there. He sees it as a plus, but the policies, the procedures, the social dynamics have rendered the U.S.–Mexico immigration situation so toxic. He's just really committed himself to making the experience of crossing more humane, making the arrival more familiar, making the going back and forth more seamless, holding a space for love and connection and family togetherness despite the realities of this border in an environment where it's become increasingly politicized.

. . .

My friends love the design firm Miller Hull. But it's confusing. A 750-million-dollar General Services Administration program with no community benefits because GSA projects don't have to have community benefits? It's LEED

The stark U.S.–Mexico border between San Diego, California, United States on the left and Tijuana, Mexico on the right.
Credit: Sargent 1st Class Gordon Hyde/Public domain.

silver, or platinum, or gold but with the traffic backing up to San Diego?[4] How can they be creating this environment of toxic air? How can we put a LEED platinum label on a building that doesn't have community benefits agreements, that doesn't have to hire local labor? And that doesn't do the environmental studies and impact analysis that it should?

David Florez, Enterprise Rose Fellow, Former Community Development Director, Casa Familiar

Connecting with this non-profit and doing small infill housing projects along a border community was the appeal of being a community architect and engaging in those next steps to help. Right away, landing in San Ysidro, my wife and I moved into an apartment that was literally 1,500 feet away from the border fence, and dove headfirst into the community planning process there. It's a community of about 30,000 residents, but this massive flow of people are moving across the border every day. In the morning, 20,000 *pedestrians alone* and 70,000 vehicles cross through the land port of entry into the United States through San Ysidro, and then the same happens in reverse in the evening as people head back home.

. . .

We were approached early on by the General Services Administration [GSA], who presented their Environmental Impact Report [EIR] that had four different alternatives for how to rebuild and expand the San Ysidro Land Port of Entry.[5] This was work also with Teddy Cruz. We secured a National Endowment for the Arts grant for the design, and we invited planners, architects, designers, traffic engineers…to come and spend three days with us to think about "How should a community like this communicate its main points on this huge project?" After the second day, the group made some very clear statements. The first was that the GSA was not thinking of the localized opportunities for really highlighting the connections between this project and the community. Second, the project, because of its size, should consider many different points of entry for pedestrians because they were the ones that were going to be the most impacted. Next, the project offered nothing for improving its environmental impacts. Finally, eventually the local economy would need to shift, and make sure that it connects to this project in a way that advances opportunities. GSA came with their Environmental Impact Report and their four proposals, but the group that we assembled had come up with criteria that we needed to hold, no matter what the design.

GSA established a local committee to review the advancement of the project. I pushed hard to make sure they understood the impact on

The PedWest bridge at the redesigned U.S.–Mexico Port of Entry/San Ysidro Transit Center. Credit: Photograph © Harry Connolly.

3.1 San Ysidro, CA

pedestrians for a community like San Ysidro. Those people shop very locally, and they're walking across the border with their purchases. I think once GSA started to look at the demolition, they realized they needed a different access point for pedestrians. So that changed their mindset and their design to make sure there would now be the existing crossing on the East, and a new entry, PedWest.

. . .

Casa is one of the first non-profits in San Diego to work with the Low-Income Housing Tax Credits and related housing partnerships. There's a portfolio of about thirteen different affordable housing partnerships that Casa has entered or is in, and then it also purchased properties that were smaller, always with the eye toward neighborhoods that have been left behind, where there's no momentum from large housing developers. It presents the opportunities to experiment.

Casa Familiar took over two community recreation facilities from the City of San Diego in the 1980s and started to just run the services. That is an older building and what is the first green space in San Ysidro, laid out around the 1900s by a utopian farming community. It's been the original central, range-hall kind of building. Then, also in the late 1980s, the organization acquired the house it was operating in. They redeveloped that to have its new social service office, and housing for individual seniors. It was the first single room occupancy (SRO) project in the community. The social service center is where Casa Familiar has always provided immigration services, translation and any kind of forms of assistance. That's always been at the center of the organization. The additional projects through the fellowship were the Casitas project, then the Living Rooms and another Estudio Teddy Cruz+Fonna Forman-senior housing project. But that last one, unfortunately, we could not move past allowance, so the organization is now looking to transform it into its first Community Land Trust [CLT] housing project.

. . .

For the Casitas project, we wanted to transform the idea of this housing project being on an alley in a way that begins to connect Casa Familiar community facilities. There are alleys in this community that are the front door to residents' homes, so we're always thinking about using design and good urban planning at the block-scale to connect A to B. By pushing the pedestrian promenade that wouldn't exist in the alley way, we augmented sidewalks into each of those properties, developed the eight homes, rotated the garages so you wouldn't see doors from the street and the sides of them could display artwork. We did special paving patterns to work with that block and had the property owners and residents build fences, and we painted those homes. We worked with artists install artwork on each of the corners.

The ability of Casa Familiar to be the investment intermediary between an entity like the City of San Diego and the community has always been a part of its work. As preparation for the Casitas project, the organization spent two years beautifying that alley, doing community clean-ups, fixing fences, painting, requesting lighting. When Casa acquired the property, the banks had

Spaces between and along Casa Familiar's Casitas in Living Rooms at the Border by Estudio Teddy Cruz+Fonna Forman.
Credit: Photograph © Harry Connolly.

been saying, "We can't lend to you, this is an alley, it's full of trash, and no, we're not going to invest our money here."

Living Rooms has ten units and two different residential buildings that are the bookends of the project. One has six units: three one-bedrooms, three two-bedrooms. The other building has four three-bedroom units. I think Casa is going try to do a twenty-five-unit project as the CLT. It's really interested in that next level of anywhere between four to twenty units, for both acquisition and new affordable housing.

3.1 San Ysidro, CA

. . .

Existing affordable housing finance mechanisms are so expensive, and they cannot reach the affordability levels the community is at. The AMI calculation is for the City of San Diego.[6] There's a gap of about 30 percent, which means the community needs housing that is about 40–50 percent AMI, and for all the financing structures, they're looking at 60, 80, or 120 percent AMI in the better-off communities. Those don't work. Casa is now looking at how to invest to secure housing for a hundred years, and this is where the CLT comes in. It's looking not only to do new projects, but to acquire existing apartment complexes that might be smaller and convert them into CLTs, and it's looking to integrate its own properties into the land trusts to be a stabilizing force in the housing market.

The second thing that Casa is doing is looking to intentionally invest in the community, with one of the first projects being a grant that fills a gap that has been on the books for like twenty-five years for a proposed ten-acre park for the community. The City of San Diego never gets enough funding to advance it. Casa is looking at doing intentional carbon sequestration landscaping or design because of the poor air quality in the community. So it's very specific about not only its mission of building affordable housing, but also how to repair a community like this.

Lisa Cuestas, Executive Director, Casa Familiar

I have been employed by Casa Familiar for twenty years, but my first year was as a volunteer helping at the annual Thanksgiving dinner where we serve about 1,200 meals, and everybody from border patrol to the fire department, to the school administrators, to elected officials would all come together.

San Ysidro residents come together at a community event hosted by Casa Familiar.
Credit: Casa Familiar.

When I moved to San Ysidro it felt a lot like home in South Tucson, where demographics are similar. But here, the border is right in somebody's backyard. This was in 2000, and you could go to lunch in Tijuana and have some great tacos, and then come back here ready to work. It's really different now, even if you have a SENTRI pass.[7]

. . .

We have a campus and a lot of spaces where we provide services or where we have arts and culture spaces. Some of these might not be solely owned by Casa Familiar, but we have been operating spaces for a long time owned by the city of San Diego. The income of a San Ysidro resident is about $45,000 where it's about $70,000 in the City of San Diego. We are annexed by the City of San Diego but thirteen miles removed. The previous CEO would joke that "They wanted the border, but they didn't want the people." What has happened historically is a lack of investment, everything from public transportation to infrastructure. At one point, San Ysidro had the highest concentration in the county of low-income housing.

The public transit, MTS, is lacking compared to what's in the rest of the City of San Diego and a huge equity issue that we've been talking about for twenty years. The San Ysidro border stop is the highest revenue generator for their entire system, but it doesn't run quick enough, so if you're a student in San Ysidro and you want to get to SDSU or UCSD, it's going to take you an hour and forty-five minutes.[8] When you talk about quality of life, that's difficult. And if you live in Tijuana and need to make that commute because there isn't enough affordable housing accessible or being produced in San Diego, good luck. That's why it's going to be interesting when we start having the density conversation, because what is allowable now on the site that's very close to the light rail could probably go as far as nine stories. But it would be the only thing that big in all of San Ysidro. The first one to try to do something like that is going to get a lot of pushback.

. . .

We had no idea what the heck New Market Tax Credits were when we were trying to figure out how to finance Living Rooms because Living Rooms isn't just housing. You can consider a lot of your social services and art spaces as commercial, and if those are generating a certain amount of revenue, then you can make it work.

. . .

One of our big infrastructure advocacy pieces is the development of a culture corridor—you might have seen the murals. To the left of that is an alley, a straight shot through the heart of the community. It's an opportunity to intersect all our leading pillars and initiatives, from advocacy and civic engagement, community organizing to education to arts and culture to direct social services and economic and community development.

Something like an alley is where you can engage with residents. We just went through this whole fruit tree planting initiative. That's how you make friends in the neighborhood. You can organize anybody around safety and cleanliness. So we do community clean-ups on that alley on a regular basis. We fund the advocacy to put pressure on MTS, the public transportation operator, to enhance the infrastructure and to pressure the mayor's office to

The *paseos* in San Ysidro are lush with vegetation and gardens and many double as more intimate scaled, social spaces. View of Cypress Drive, the future cultural corridor that is envisioned to further stitch the neighborhood together with additional permeable (green) infrastructure, art and other amenities.

deem the culture corridor a "sexy street." That's his term. But that means we must get a plan together to keep pedestrians safe, slow vehicles down, and make sure that arts and culture continue to be infused in that space with green infrastructure.

Sometimes San Ysidro doesn't even come up on planning things because they just say "Border Community." There are folks that live here half the week and in Tijuana the other half. But there are also people that have lived here for thirty, forty, fifty years, and they don't think badly of the border, but that is just something they deal with the impacts of. They almost never cross the border because they don't really have to, or when they do, they do it for recreation or to go visit family, not because work or home is on either side. But, the closer you get to the border the businesses cater to the border crosser. West San Ysidro Boulevard, which is close to the culture corridor and our other properties, is just now starting to resemble something that looks like it caters to the community that lives here and not the daily border commuter.

A Conversation

Teddy Cruz (TC) and Fonna Forman (FF), Principals,
Estudio Teddy Cruz+Fonna Forman

TC: We are more than designers of things, we're designers of civic and economic processes. When we began this project, we wanted to argue, at least for the field of architecture, that instead of waiting for the client we could imbed a project through a series of relations that were well-rounded. To stitch

View over San Ysidro with Living Rooms at the Border (center of frame) and Paseo la Paz, new affordable transit-oriented housing in the distance. Credit: Casa Familiar.

together a research agenda, so it's not just the architect coming in to design a project. It's also about building trust in the long term.

This project began with a very close relationship with the previous director of Casa Familiar, who is kind of the matriarch of San Ysidro, Andrea Skorepa, with whom we built an incredible trust throughout the years.

It is a long-term commitment. Our practice has been embedded in not only that neighborhood but also across the border, and began with a lot of assets within the community, but a very precarious economic foundation. This begins a way for us to critique conventional approaches to urban planning. Affordable housing developers are still imbedded in very conventional logics of economic development. We wanted to prove that something else could happen here that didn't have to rely on tax credit-based structures or on the logics of certain policy-oriented processes.

Part of the challenge when we began working with Andrea was, could we rethink the model of development altogether from the bottom-up so that communities become developers themselves and they don't have to wait for that affordable housing developer with the typical tools—non-profits who qualify for tax credits but remain only as service providers and not the owners and the managers of their own modes of productivity? **We asked, can neighborhoods be developers?**

FF: It's important to understand how much social capital already existed in that community. Casa Familiar is a deeply rooted social service agency that's been around for nearly fifty years, and they have a whole array of social service programming and had really built-up trust over time with this community. The nonprofit had aspirations to build housing, and they were already serving as a

3.1 San Ysidro, CA

local city hall that would enable residents to achieve certain kinds of ends. So that trust was already in place. It was a question then of how to align their aspirations as a community with all the restrictions and zoning obstacles that prevented what they envisioned for their own neighborhood. Over time, the university and the community became partners in pushing against these obstacles.

There are all these assumptions about allowing communities to self-determine, but there were policy avenues and zoning avenues that the community was not aware of. Similarly, there was so much knowledge in the community that we needed.

TC: It took time for us to realize that we didn't want to be the architects that just come and listen to the community then go back to the studio to design. No, we wanted to focus on tackling and challenging certain reductive notions of advocacy planning. It really enabled a process of exchange with the community, where the push and pull constructs knowledge together. It's an interesting case of the meeting of knowledges between us architects and communities that eventually becomes a story of a cross-sectoral collaborative model that invests in the small scale of development.

We would have nights with Andrea at times discussing issues with her associates. For example, at some point, Andrea had led a moratorium on development in her own community because she was afraid developers would install the typical recipes of beautification and produce gentrification. They were afraid of density. **But we were saying that part of the problem is that density has been misunderstood. It's not about size; it's about the intensity of the social interactions.**

We discovered that in the 1990s, in the moment of the ascendance of neoliberal economic policies and political agendas, booming construction made downtown San Diego incredibly wealthy, with luxury condos everywhere. At a moment of urban explosion and development, not one single affordable housing unit had been built in many of these neighborhoods. Part of the problem is that for an affordable housing developer to be competitive for tax credits, the building would have to be at least fifty units in density. Well, that's prohibited in places like San Ysidro. Obviously, developers would not touch it. So part of the idea was, how do we begin to rethink the model? Our work researched informal urbanization across the border, the positive impact of immigrants in transforming neighborhoods, the leadership of non-profit organizations that effectively become informal city halls. All these amounted to this understanding: *communities could develop their own housing.*

. . .

Fonna and I discovered that the economic power of the university could in fact be the last layer to make the whole sandwich of capabilities and resources the development needed. Through the university agenda, which incrementally began to build in collaboration with Casa Familiar, there was an opportunity to invest in the public space.

FF: Through the university we were able to secure funding for programmatic activities, and we were able to raise capital funds from philanthropies and ArtPlace America.[9] We were able to get the social seed of the project completed first, and it was then that we were able to leverage those cultural amenities to qualify for New Market Tax Credits and to assemble a development package for the housing. It sort of turned the model inside out. Instead of developing the housing and finding a way to squeeze in social amenities, it was a deep investment in the social and cultural amenities that qualified the non-profit for housing subsidies and funding.

. . .

FF: When architects develop their own properties, they recoup their investment later on through profits. In this case, our salaries are paid by the university, so we're not ever seeking to profit personally from these projects, but investing our time is a way of getting the projects off the ground.

TC: It was a larger premise that nothing should prevent architects from developing their own projects through collaborative models. A notion that we have been very interested in advancing theoretically is the bundling of resources. The horizon of accessibility to the development of the city is always controlled by the same actors. We need to expand that.

. . .

For this project, through some back doors and loopholes, we were able to maintain the community theater by disguising it as a kind of support system for the housing, and the accessory units were presented as storages. Ideally, as a prototype, it could potentially enable the city to be more open to the fact that we need to reimagine density and they might be more willing to transform zoning. We wanted to prove that zoning could be generative to organized, local activity and the local economy.

. . .

In one of our workshops we had a game with the board reflecting the parcels of the block in San Ysidro and the alleys and the streets, pieces of wood that represented units, little trees, cars. I said, "What would be the first thing you'd do?" This lady, who was very old, took a piece of wood and put it on one parcel, and put another on another parcel, and she basically built the reality of her own block. Then, somebody else said "What would happen if we put the house towards the alley so there is a space in between, and we can put another little unit there." And all of a sudden, the board was full of wood. The lady said, "When I was a teenager in Guadalajara, Mexico, my block wasn't made of individual buildings. I would open a door to my house, and I would go through a corridor and find a courtyard." From there, we began to talk about neighborhood collaboration. She said at the end of the workshop, "I cannot believe these houses are so selfish," and that dropped it for us. That breakthrough opened the parcel in the way that it became design: small-scale development, connecting alleys to streets.

The small white buildings that contain Casa Familiar's social services are a model of an accessory building. An accessory building, if it is no larger than 10 x 12 feet (3m x 3.6m) (and if we give it 15 feet (4.5 m) ceiling heights, and if we give it a big window and utility sink) could be a beautiful artist studio, and would not require a permit, according to local code. Which meant, *can the community build it?*

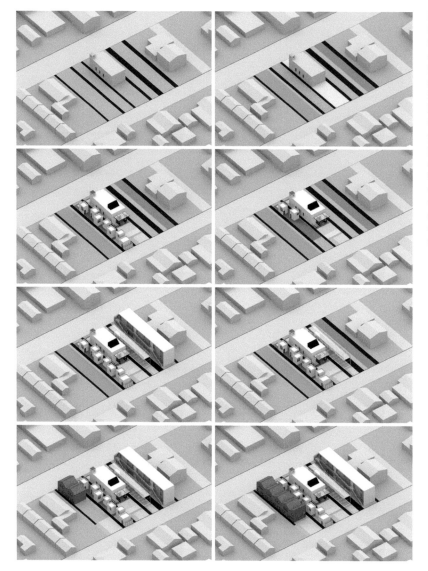

Phasing diagrams by Estudio Teddy Cruz+Fonna Forman created to illustrate their ideas for "socializing small-scale urban fabrics in suburban neighborhoods." Designs looked to advance scalable prototypes based on the principle "that community-led, incremental small-scale development democratizes urbanization" (Cruz and Forman, 2022). Credit: Estudio Teddy Cruz+Fonna Forman.

. . .

TC: Fonna directs the Center on Global Justice at UCSD, where I direct the cross-border urban research. And the Center on Global Justice that Fonna was able to construct is an interesting platform within the University of California system, which allows levels of autonomy. We don't report to deans. We report directly to the chancellor. It is a center that is focused on opening new agendas. This had been a long process of designing and calibrating these institutional protocols to not only be more relevant in connecting top-down resources to the bottom-up intelligence, but also to bring the intelligence of the communities to teach inside the university. We were arguing the social, political and economic knowledges of activists, like Andrea Skorepa, could reorganize our research agendas inside the university, and that's the reason we were given support by the Andrew W. Mellon Foundation. Now we're negotiating the renewal for the next five years because of the success of the Community Stations.

FF: If we didn't have the kind of institutional agility that the Center on Global Justice provides within the university, we could never do this kind of work. We used this global center to go local into border communities because after all, *global always happens somewhere, right?* And we live on this border which is an amazing international laboratory.

Craig Curtis, Partner, Mithun; Former Design Architect for the Border Crossing Project with Miller Hull

We had to keep the entire facility open 24/7, 365 as we built a 750-million-dollar project on top of the existing operation. We needed to try to keep that pedestrian experience at grade and make it feel like you're really just walking into the country, not going up this concrete ramp and into this giant hall that's filled with security and then back down. We weren't going to win it in any kind of a beauty contest. We just had some good ideas.

We were working with two primary clients, and I saw them as equal. GSA was paying the bill, and running the project, but the main tenant was Customs and Border Protection (CBP). We were so fortunate we had Maria Ciprazo from GSA and Oscar Preciado from CBP. I just clicked with both of them and we couldn't have accomplished the project without them.

After we got the project, we were working away, and I asked Oscar one day, "How in the world did we ever win this project?" We were the smallest firm going after it. And he said, "You were the only lead designer who actually took the time to go walk through the border and back and experience what it was like." I think that says a lot about Oscar's approach to this, too, that it wasn't just all about security. That border crossing had been there for decades, so people were really used to it. They exited out of that little plaza built up with small shops. There was a little taco shop there that was fantastic. That one little retail strip that had grown over time was the only thing we really had to work with, and the best thing we could do was preserve it, embrace it, bring people out to that area, and hope that it would continue to thrive.

There was enough budget to do some design moves that were not inexpensive but made a big difference, architecturally. The entire entry canopy for the vehicular experience I feel good about. Most borders you come up and feel like you are just waiting in a giant parking lot in front of this row of toll booths that are not really that interesting to drive through. We've got this dynamic canopy with a public art piece on the leading edge done by Höweler + Yoon Architects. The canopy is, I don't know, a thousand feet (305 m) long. It's a remarkable structure to provide shade.

It's not only about providing a dignified experience for the Mexican population coming into the border, but also the people who work there. These border crossings can be hot and uncomfortable and dusty, and that can get to your psyche as you're trying to do your job. We tried to soften all of that. We used a translucent fabric that has this wonderful, filtered light coming through, so it's not just these giant canopies that are dark underneath.

View of the rebuilt $750 million San Ysidro Land Port of Entry. Dynamic and ever evolving, the southern border between the United States and Mexico is anything but ambiguous.
Credit: Chipper Hatter.

I think you can lower people's heart rates and things if you're in a place that has access to light and is an interesting, uplifting architectural experience. They were very supportive of that, even CBP. They saw it was worth the investment to make some of these moves. It hadn't been done before on a southern border.

We had pretty strong sustainability goals going in. It's not able to hit net zero right now, but with some off-site solar it can get there. I'm also really proud of that fact that water is as important as energy down there. We have a black water system on site, the first time this has ever been done in a GSA facility like this. We capture, and treat, all the sewers on site and use that for irrigation. The drought situation in California is terrible. The southern border is really dry, and anything we can do with water conservation is something that we were really striving for.

Notes

1. Of the fifty-two crossings, eight are rail lines, forty-three are roadways (twenty-four bridges, two dams and seventeen roads), and one is a ferry (Rose and Davidson, 2010).
2. An accessory dwelling unit (ADU) is a small additional (residential) building that someone builds on their own property in addition to the main house.
3. The state of California is simultaneously investing in light rail, heavy rail, high-speed rail, bus rapid transit and active transportation.

4. LEED stands for Leadership in Energy and Environmental Design, a green building rating system by the U.S. Green Building Council.
5. An EIR is an environmental impact report containing an overview of a project and its potential impacts. The document describes measures to reduce or avoid adverse impacts as well as recommend alternative courses of action.
6. AMI is the Area Media Income, a standard for determining income eligibility.
7. A SENTRI pass is the Secure Electronic Network for Travelers Rapid Inspection, a U.S. Customs and Border Protection (CBP) program that allows expedited clearance for pre-approved, low-risk travelers upon arrival in the United States. Participants may enter the United States by using dedicated primary lanes into the United States at Southern land border ports.
8. San Diego State University and University of California San Diego.
9. ArtPlace America was a ten-year, $150 million cultural initiative that brought together foundations, federal agencies and institutions to advance issues of equity in urban planning.

References

Bruns, R. (2019). *Border Towns and Border Crossings: A History of the U.S.-Mexico Divide.* Santa Barbara: Greenwood.

Casa Familiar. (2023). Welcome; Mission. Retrieved from https://www.casafamiliar.org/

County of San Diego, Health and Human Services Agency (2022). *Poverty in San Diego County: Areas of Concentrated Poverty and Food Insecurity.* County of San Diego, Health and Human Services Agency.

Cruz, T. and Forman, F. (2022). *Top Down Bottom Up: Socializing Architecture.* Berlin and Cambridge, MA: Hatja Cantz Verlag and MIT Press.

Forman, F. and Cruz, T. (2021, February 22). David Gamble and Patty Heyda, interviewers.

Forman, F. and Cruz, T. (2024). Foreword: Radical Mapping: Critical Proximity. In Patty Heyda, *Radical Atlas of Ferguson USA*, Lakewood: Belt Publishing.

Rose, A. and Davidson, D.L. (2010). *Atlas of Land Entry Ports on the US-Mexico Border.* Border Policy Research Institute. Retrieved from https://cedar.wwu.edu/bpri_publications/30?utm_source=cedar.wwu.edu%2Fbpri_publications%2F30&utm_medium=PDF&utm_campaign=PDFCoverPages

Stockard, J. (2023, September 28). David Gamble, interviewer.

Swenson, K. (2020). *Design with Love: At Home in America*. Atglen: Schiffer Publishers.

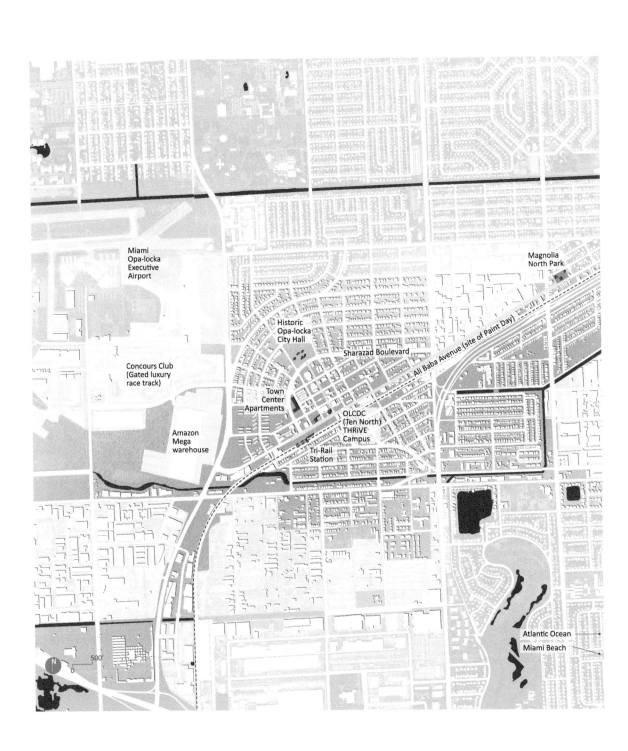

3.2
Opa-locka, FL

Opa-locka, Florida is a small town in the shadow of the city of Miami just 11 miles (17.7km) northwest of the downtown and beaches. It is what we classify as a *metropost*, since it sits within a major city-region among thirty-three municipalities in Miami Dade County (2021 pop. 2.66 million), and is part of the large metropolitan area surrounding the City of Miami (2021 pop. 439,890) (United States Census Bureau, n.d.).

Opa-locka was created in 1926 as an Arabian Nights-inspired destination to complement other themed districts in and around Miami at the time. Home to 15,373 people, the majority of whom are Black and Hispanic residents, Opa-locka in 2023 remains the site of one of the largest concentrations of historic Moorish architecture in the United States (World Population Review, 2023).

Yet the metro region that Opa-locka resides within has the second-highest rate of income inequality in the United States (Florida and Pedigo, 2019). This means that while the region ranks tenth in the world for the most billionaires among full-time residents, over 40 percent of Opa-lockans live below the poverty line (United States Census Bureau, n.d.). An illustration of this stark disparity lies at the edge of Opa-locka itself, where an Executive Airport serves private jets and elite business interests. Here, a racetrack for luxury automobiles and a massive Amazon warehouse sit adjacent to a town still sorely underfunded. To complicate these extremes, the small municipality has infrastructure unable to address impacts of global warming, and flooding is a common occurrence.

Despite these challenges, or perhaps because of them, Opa-locka has been home to a cherished long-standing community development corporation. The Opa-locka Community Development Corporation (OLCDC; renamed Ten North in 2022) has provided over forty years of consistency and stability, alongside a steady rebuilding framework that leverages the arts and culture of the town and its people. **Through its CDC, Opa-locka expands the narrative of economic development capital toward rebuilding that prioritizes** *community and cultural capital*.

Scales of (Shared) Power/Planning

OLCDC was founded in 1980 "to transform lives by providing equitable access to housing and wealth-building opportunities" for residents of the under-resourced community, "particularly those affected by systemic bias" (Ten North Group, 2023). Over the years, the CDC engaged in various projects that challenged politics as usual by investing in low income housing and social services, and by

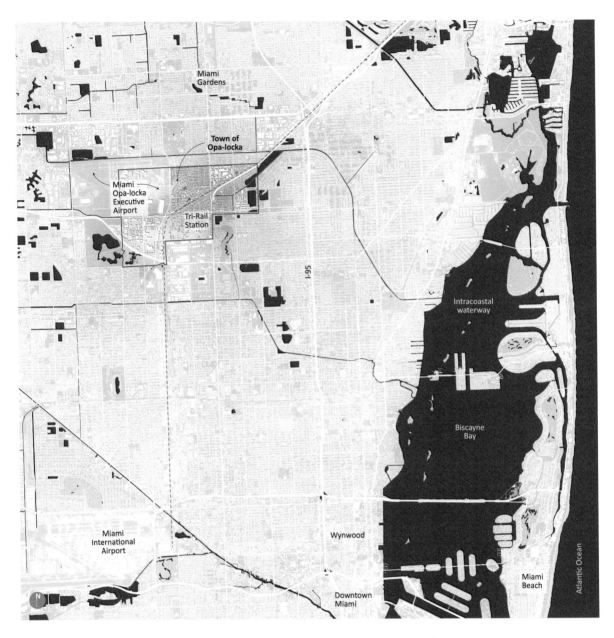

re-centering design, to shift value narratives and stage change via the arts. The vehicles for such work were a series of national level grants that supported neighborhood engagement. These grants included a Housing and Urban Development (HUD) Neighborhood Stabilization Program grant in 2009 and a 2012 HUD Choice Neighborhoods grant. An additional National Endowment for the Arts (NEA) grant in 2013 opened more creative thinking around neighborhood challenges. The earlier grants helped stabilize affordable housing in Opa-locka's neighborhoods, and the NEA grant allowed for two notable design-forward initiatives, *Made in Opa-locka* and *Oasis/Oases*. These efforts brought designers in from across the country to test new ways of creatively reclaiming the histories and possibilities of place with smaller-scaled urban interventions.

Moorish-Revival buildings and features characterize civic space in Opa-locka, Florida. The town was speculatively developed by Glen Curtiss in 1926 and twenty buildings in this small municipality are listed on the National Register of Historic Places (The City of Opa-locka Florida, n.d.).

Made in Opa-locka was led by young architects: Germane Barnes from Chicago and Los Angeles, Jennifer Bonner from Boston, and Christian Stayner from Los Angeles. Their ideas included incremental insertions that, for example, transformed an existing church into a performance venue, and reused an abandoned house for civic purposes, plus other concepts described as "micro-enterprise zones" to spur local development (Pacheco, 2017). However, it became clear that lasting work would require a more sustained designer-city-commitment. Germane relocated permanently to Opa-locka as the community's first "designer in residence" at the CDC to build on the initial efforts. *Made in Opa-locka* grew its projects and capacities.

With a designer in residence, innovative urban ideas could be actualized with partnerships, buy-in and support for projects that required longer timeframes or more local connections to mature. From 2013 to 2017, *Made in Opa-locka* realized projects including the Magnolia North Community Park and elements of what became known as the Thrive Innovation District: "A compilation of services and resources located within the City of Opa-locka … designed to assist the community to become financially self-sufficient and empowered through education, workforce development, health initiatives and entrepreneurial opportunities" (Barnes, 2021). The Thrive district is anchored by the ARC, an arts and recreation center with a gallery space; an urban farm, and venue and office spaces at the historic train station.

3.2 Opa-locka, FL

Meanwhile, the CDC commissioned Los Angeles-based landscape designer Walter Hood to rethink Opa-locka's Ali Baba Avenue, a wide unshaded street that was prone to flooding during large rain events—a structural problem that contributed to school absences for kids unable to traverse water-filled routes to school. This collaboration became known as the 2013 *Oasis/Oases* series of proposals that included shade structures, vegetation and beautification elements to enhance walkability for pedestrians. Hood and his team conducted a series of design engagements with residents, the CDC and community partners to develop the interventions including infrastructures to drain and manage flood waters. But funding for implementation of the larger scheme was tangled in political webs at local and county levels. So while they waited for funding, the collaborators staged another more visible ground-up design project. "Paint day" brought the community together for one day to create a dynamic pattern of bold graphic designs that were able to literally reclaim the neglected street in real time. This maneuver demonstrated immediate commitment with immediate results, and it brought civic leaders together with residents and other partners. The design evoked the vibrant colors and patterns of Moorish arts.

It can be a challenge to foster a sense of campus in urban environments where buildings and programs are spatially disparate. More than just providing a venue for engagement, the tactical urban effort reinforced the close proximity of the CDC's buildings to one another along Ali Baba. Rendering the ground plane demonstrated that the distance between them was closer than it appeared.

Germane Barnes credits the community for the design and implementation of Magnolia North Park that he helped bring to fruition during his designer-in-residency at the OLCDC in Opa-locka. He says, "We didn't do a single thing without community buy-in. The neighborhood kids designed the new Magnolia North Park themselves" (Carnick, 2022).
Credit: Germane Barnes/OLCDC.

"Painting Day" on Ali Baba Avenue in Opa-locka, Florida. Credit: Hood Design Studio.

Reclaiming "Creative Class"

"Place-making" is over-used in urban design today as a catch-all term for marketized projects that serve consumers and tourists more than residents. Opa-locka "makes place" on its own terms, by controlling the narrative amidst narrow (and incompatible) accumulation-driven mandates that don't pay attention to local meaning or needs.

Opa-locka's revitalization has been dubbed a "Wynwood-based strategy," after the Miami industrial district that became a global destination for contemporary art. While the CDC does not seek the gentrification that happened to Wynwood, it has tapped into the role of art to tell important stories and define place. The CDC embraced an urban rebuilding strategy centered on the arts, but their effort stands out for its focus more squarely on *African and African American* cultural production. The ARC hosts rotating exhibits of emerging Black artists, and in 2022, Opa-locka became a site for the famous global Miami Art Week, anchored by the international art fair known as Art Basel.

Richard Florida famously coined the "creative class" in 2002 to describe the American urban economic sector that capitalizes on trendy makers and entrepreneurs (Florida, 2002). Critiques of the model have pointed out that most of the "creative" vanguards ushered in sweeping gentrification that seldom helped existing residents in the parts of town they occupied. Opa-locka rescripts "creative class" away from gentrification for capital markets and toward a notion of creativity to mark existing space and lift community capacity. Anchoring creativity to regenerate the town has led to a vital re-centering of African American traditions and art. The ARC has shown the work of globally renowned artists such as Ebony G. Patterson, Dred Scott, Olalekan

Jeyifous, Hank Willis Thomas, and many others. Before the exhibits and Art Basel connections, Olalekan Jeyifous collaborated with the tactical design team to paint murals on the exposed walls of the CDC campus as it came together. These murals, like the paint day project, evoke the colors and patterns of Moorish culture, but also convey a distinctly local set of ideas about community in Opa-locka.

Designing a Process

What does it mean to practice urban design in a community perpetually disenfranchised by structural inequality? Design becomes less about theories and rhetorical representations and more about its creative capacity to communicate and coordinate a project. Plans themselves—as aesthetic, promotional tools—are cast aside for processual tools to initiate and complete the project as well as engage and inspire the community. Germane Barnes summarizes:

> A lot of these things weren't about design, they were about community need. I was like, I'd rather just go ahead and do the work. … The engagement is what was more important—way more important than me coming up with drawings. More time went towards that than it did

The ARC in Opa-locka centers Black designers and heritage in every facet of rebuilding: The mural on the building is designed by artist Olalekan Jeyifous, as part of the urban street art series of projects also in collaboration with landscape designers Hood Design Studios. Exhibits regularly showcase African and African American art and artists (VenusInOrbit, 2016).

> to the actual sort of diagrammatic way that we show work, which is more like "here's the construction documents, let's build this thing, [and] then let's move on to the next."
>
> (Barnes, 2020)

Urban design in Opa-locka doesn't have a formal "master-plan," but what one finds as a plan are the lines of input from people who live and work there, and the projects they've brought to fruition with the CDC. Those projects are slowly defining a central cluster of buildings and spaces that feel like a distributed campus. The term "community engagement" is commonly used in design processes without clarity or critical reflection of who the "community" really is. Opa-locka's story of gradual rebuilding is a very specific *project of relationships*; of designers talking to, listening with—living with—and learning alongside and from the public. As his residency concluded, Barnes went on to become a member of the Opa-locka city council, so he could continue his planning involvement via zoning and code changes and other structural systems that underlie how and for whom redevelopment happens there.

Stakeholders

Similar to place-making, the term "stakeholders" is clichéd and overused in urban design and planning endeavors suggesting, incorrectly, that each player has somewhat equal weight in decision-making processes. Planning with stakeholder engagement in Opa-locka is not without major corporate power brokers in view. Ringing the town is the Opa-locka Executive Airport, controlled by the county. "Bill Gates has a hangar there" (Barnes, 2020). Business elites fly in and drive immediately out much like the airport's related developments and spending that skip over the local municipality as well. In 2019, the multi-national Amazon Company added a massive fulfillment center to the airport complex. Bringing few jobs or substantial changes to the area, the immense structure only reinforced the industrial chasm between the town and its primary transportation corridors. That deal was negotiated by the City of Miami and Miami Dade County via the airport, bypassing Opa-locka local government (Barnes, 2020). Yet the Amazon Company brought employee volunteers to help complete and paint the final fence around the CDC's *Made in Opa-locka* Magnolia North Park project (OLCDC, 2019). Gestures like this are valued between town and company. But they also signal the unspoken contract that persists between uneven 'stakeholders' who, despite the idea of the term, don't actually all hold the same access to power, decision making and benefits.

Amazon Inc. is a global empire and one of the most recognizable brands in the world. Similarly to the Bentonville case in this book and the presence of the Walmart corporation headquarters there, Amazon in Opa-locka holds vastly different access to capital than the town or CDC does. On the one hand, formidable companies can override decision-making in weak market neighborhoods to exploit cheap land and steer policy in their favor. On the other hand, cities with serious fiscal deficiencies rely on the philanthropic largesse of its major corporations—and jobs—even if the company's activities might undermine the town's long-term environmental and financial health.

A Moorish-styled building houses the Opa-locka CDC's offices in the center of town near the ARC, the train station and other parts of the CDC's growing Thrive Innovation District campus.

Apart from corporate actors, the CDC holds its own to maneuver redevelopment at times when the town can't. It is notable that key leaders, like Vice Mayor Chris Davis who was elected in the wake of a 2016 political scandal, came directly from community organizing work with the CDC. Because the CDC has been in town for so long, there is a fluid, reciprocal relationship between it and the government. The community development organization can do things the city can't and the city can step in and try things the CDC couldn't do. Outstanding issues can be taken up again by the same person in a different position years later. Chris Davis tells a story of how, as an elected leader, he could go back to aid a housing project that he had once advocated for as a community organizer. Inversely, Dr. Willie Logan, the executive director of the Opa-locka CDC (Ten North) had been Mayor of Opa-locka years prior. Some see this continuity as discouraging to new elected leaders hoping to shape their own agenda. Nonetheless, these dynamics have as much to do with the small size of the town that allows the same people to move between different public or private sector positions as it has to do with the persistence and longitudinal engagement of committed actors and organizations.

CDCs fill gaps left by American political economic austerity (and privatization) that has left local governments stripped of capacity to provide services or lead major rebuilding efforts. In weak market cities and towns, there is low development confidence, making it hard to achieve the public–private partnerships meant to fill the direct public funding voids. In this predicament, CDCs are all the more crucial to neighborhood stabilization and rebuilding. Not only do they provide wrap-around services for the general welfare and support of people in the community, but they also frequently function as a developer that creates its own amenities and property markets where they would otherwise still be waiting for them. See also Casa Familiar, another effective CDC in San Ysidro, California in this volume.

Like so many towns across the United States, Opa-locka is a site of racialized extraction when tools of capital redevelopment and finance are

The Town Center Apartments, a new affordable housing project in Opa-locka, Florida anchored by the CDC-led contemporary sculpture installation by Hank Willis Thomas titled *All Power to All People*.

3.2 Opa-locka, FL

abused. The town's poverty was leveraged to qualify the executive airport and Amazon development areas for "Enterprise Zone" incentive status, an overlay that suspends certain government regulations or taxes for qualifying industries in underserved areas. The tool benefited the corporations there but has not necessarily improved life for residents, even though many community development agencies are still compelled to participate in these kinds of financial markets to survive.

The CDC in Opa-locka operates as its own developer of affordable housing, and their real estate investments help fund the vast social services they provide. As the county and region offer redevelopment tools to place lucrative industries, or bring higher-income projects and residents (at risk of gentrifying the town), the CDC must draw from these same tools, or structure new ways to self-determine development activities. In the face of this increasingly uneven power and scalar balance, the Opa-locka CDC recently changed its name to Ten North, to capture the larger radius of other towns around Opa-locka that they had already been serving. This gesture creates the optics of a more unified stance among the many smaller towns, to better face regional and corporate power politics.

Rebuilding in Opa-locka is characterized most by the innovative ways its Community Development Center re-claims and holds space for people when the structural systems of American rebuilding persistently fall short.

Development Climate

It may seem unlikely, but against the backdrop of disinvestment in Opa-locka, there remains the risk of gentrification and displacement, as sea level rise and housing costs in Miami Dade County make living on the coast and in downtown Miami increasingly unattainable for the average person. For the moment, Opa-locka may be safe from wholesale capital market gentrification because of its peripheral status to downtown Miami (and the beaches). But it is increasingly at risk of "climate gentrification," as the greater Miami-Dade County population continues to grow and moves inland because of recurring coastal flooding.

While the OLCDC's activities include crucial social service offerings to support residents, their work as land-bankers and affordable housing developers constructs a future buffer against these threats. The CDC has made market-savvy land acquisitions that allow them to hold strategic parcels in anticipation of encroaching gentrification, all while elevating design and testing new forms of spatial agency. The organization's projects span a number of blocks and buildings along Ali Baba Avenue and into the neighborhoods, establishing a grassroots imprint across the town and in other towns around it.

If economic connections to downtown Miami are tenuous, or contested, Opa-locka is fortunate to have an enhanced *physical connection*. The South Florida Regional Transportation Authority, or Tri-Rail, provides frequent train service between West Palm Beach, downtown Miami and the Miami International Airport, a distance of 73.5 miles (100km) with fifty weekday trains. The Opa-locka station occupies a central site directly across from the

A flooded street in Opa-locka near the airport after a brief but heavy downpour in 2022. Basic street and sewer infrastructure no longer has the capacity to address frequent rain events.

CDC's administrative offices. The presence of a regional commuter train system and station in Opa-locka poises the town for new residents and commuters, and there are numerous vacant and underutilized sites in the vicinity of the station which signal opportunities for higher density, affordable transit-oriented development. Still, car culture reigns. The streets are wide and sidewalks are in disrepair. Bikes have become a positive mobility stand-in among many residents without access to cars. At the same time, routine flooding from increased rain events make biking and walking around town a challenge.

Walter Hood's *Oasis/Oases* project had addressed the streetscape in Opa-locka, not just for aesthetics and walkability, but to prepare for the persistent flooding that global climate change was amplifying in this tropical context. Yet funding never came, reinforcing the American urban redevelopment

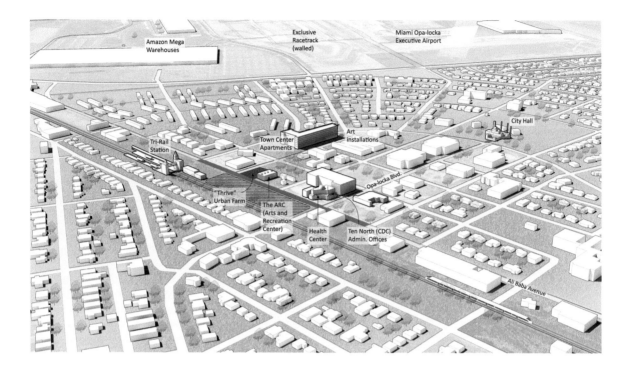

paradigm that strictly ties public investment to capital markets in order to manifest needed infrastructure changes. As sprawl continues on greenfield sites outside of towns, or as wealthier neighborhoods receive incentives to further capitalize, the areas that need investment the most are often the least likely to receive it. This has pronounced impacts in a system also defined by cuts to social welfare and public space.

Perspectives

Germane Barnes, Former Opa-locka Artist in Residence and Opa-locka Planning and Zoning Board Member; Architect at STUDIO Barnes; Associate Professor and Director, Community Housing & Identity Lab, University of Miami

The Moorish architecture was purely a tourism and marketing ploy when Opa-locka was first settled by Glenn Curtiss. The idea was that other cities, such as Coral Gables, which is Spanish revival, and then Miami Beach, which is art deco, had some sort of architectural character to them to make you want to visit. And so Opa-locka could have a Moorish, Alibaba, Arabian Nights thing. But he ran out of money and then there was a massive hurricane. So only a third of the original buildings were actually built. It never reached the capacity that would make it a tourist attraction, though you still have the largest collection of Moorish Revival architecture in the Western Hemisphere.

In the last five years, the city has been more open to changing its zoning and LDRs (Land Development Regulations) so that it's not as restrictive. There were so many Moorish restrictions within the code that really stifled any innovative or contemporary contributions. Graphics and patterns didn't get allowed till 2015. Prior to that it was solids and color blocking, but then when we did our arts and recreation center, we were supposed to get a temporary, thirty-day stay to put up this graphic mural and we just disobeyed the ordinance and extended it. The city ended up actually liking it, so they put forth an amendment to the city ordinance so that you could have those things in the code.

. . .

Our projects were very small incisions that I would define as tactical urbanism as we hear about, but aside from that, it's really just about being community-oriented, with community-directed design. Because at a certain point in order to do these types of projects, you have to be willing to remove the architectural ego of aesthetics and go down to the bare necessities of: *this is what people are asking for, this is what I have to give them, this is the budget that I have to work with.*

The people I work with are really interesting and cool. They've been lifelong residents of Opa-locka, they just don't have any formal architectural or urban training. I help them out with that so they're doing what's best for their neighborhoods. Because they have all of the wisdom and the insights that I could never have, not being a native.

Some of Opa-locka's National Historic Register-listed buildings have been restored. Others, including the massive city hall complex here, are at risk of deterioration as they await funding for rehabilitation.

Chris Davis, Opa-locka Vice Mayor/City Commissioner (2018–2022), Former Community Organizer, Opa-locka Community Development Corporation; School Teacher

In addition to all the things I do for the city, I'm also a second-grade reading and math teacher.

I grew up here in Opa-locka. My first job out of college was working for the CDC for several years, mostly as an organizer. I got to see things from the city's perspective from a grassroots point of view. And in that role, I lead several community initiatives. One was advocating for a notorious housing complex in the city of Opa-locka, called Glorietta Gardens. I was putting together a tenant association there, which eventually ended up leading the residents to advocacy and forcing the property managers to make much-needed improvements to the complex.

In maybe 2010 or so we received the grant from the federal government called the Neighborhood Stabilization Program (NSP). NSP was a cluster of federal dollars aimed at stabilizing property within urban neighborhoods after the housing market crashed in 2008. NSP was specifically aimed at buying single family homes, so we used that to target certain residential neighborhoods in the city. We got them at really good prices because the market was low, and we were able to turn those properties around and put them back on the market and also work with the first-time homebuyers' program to counsel our community and ultimately get them mortgage-ready.

. . .

One of my first tasks when I was elected Vice Mayor in 2018 in the city of Opa-locka, was to go to Glorietta Gardens where I had started the tenant association and say hey, remember me? Because of course now I had power, I could bring the city to the table to ultimately resolve many of the issues that I tried to resolve as a community organizer. I think that was my biggest aha moment.

The walled conditions around the Opa-locka Executive Airport (at left) and one of its gated executive clubs, a luxury racetrack complex (at right).

I worked for non-profits, I've written grants, I've done several needs assessments, so I know all of our deficits, but I guess part of my journey as an elected official was learning how to leverage those deficits. We are the poorest city in Miami-Dade County, but also where you see challenges, there's also opportunity. And I think as an organizer, I've always kind of embodied that school of thought. It ultimately challenges you to change your perspective of how you see community.

. . .

We have our own private executive airport that is an Enterprise Zone, which is a tool for development. The airport gets all the tax credit, all the subsidies, as far as development goes, and they use the adjacent neighborhoods' tax base—that of the poorest city in Miami-Dade County [to qualify]. The whole purpose of an enterprise zone is to stimulate the local economy, to bring businesses, to create jobs. It's supposed to benefit the residents. But I'm pretty sure, just like other enterprise zones here in Miami, you don't see a lot of benefit to the urban core, which is the original intent.

About two years ago, we actually had Amazon come in, get all the tax credits, and get a place at our airport site. Our airport is a private entity within a city that Miami-Dade County actually controls, so Miami-Dade County axed us out of the whole arrangement. When the facility was built, our city didn't receive any of the impact fees typical for any new development in any city, but given the size of Amazon, those impact fees would've been significant. We don't provide any services to Amazon nor have we received any direct tax benefit. We did receive one indirect benefit with them being there: In the first year or two since they've come into our community, we've seen a growth in our tax base, which benefits our community in many ways. The dilemma is, they're in our community and even in our water service district. So Amazon is definitely one of those battles that we need to fight and address. And even the airport, they're building high-end services. They have a high-speed racetrack that is really expensive, that nobody in our community would patronize because it's not affordable. So when you see those type of decisions being made and development done in that way and your county supporting it, it makes you think, like "Wow, things aren't as clear cut."

. . .

We had a saggy pants ordinance, where prior officials thought that saggy pants—[worn by] young African American and Hispanic men and women—were such a problem that our city decided to adopt a law. Even as a resident I was always opposed to that law. I always said if I ever had any power, I would repeal that law the first chance I got. Well that opportunity came when I was elected in 2018. And even during the pandemic, we ended up repealing it, right around the time when the country was protesting the George Floyd incident. The goal was to make our city more equitable and provide fairer policies. And it ignited a national conversation that I can't say I was ready for. But it made all the national news networks, USA Today, CNN, Fox News.

Walter Hood, Creative Director and Founder, Hood Design Studio, and Artist

I became fascinated by this central spine and the community. The street—that was about a mile long—connected the downtown where the governmental offices were, with where people lived. And along the street was the most barren landscape I've ever seen, particularly in Florida. It was in August, it was so freaking hot. I was walking down the street, and people thought I was crazy that I was walking! People in their car asked me, "Are you OK?" I asked Willie [Logan, CEO of OLCDC] if that could be my project. And that's how we began our work there.

The temporary work we ended up doing was not the "project." We came up with a one-mile concept for the street landscape based on the idea that they didn't have enough money to improve the entire street. The sidewalks were inadequate, like thirty-six inches wide, there were hydrants placed there. So from an ADA perspective, even just the simplest thing would have cost a lot of money. Two, drainage is a huge issue there because of these flash storms. Literally, when it would rain, the streets would flood. And we found out that rain was the number one hindrance for kids' absenteeism in school. So we said, "OK. What if we made these oases along the street?" Which would be—instead of making one linear thing, you did these patches, maybe each one happening an eighth of a mile, so that people as they're walking, they would have these places. And then we tried to tie that with urban infrastructure, taking the curbing, gutter, things like inlets, and making those the basic design. Through using the sidewalk and the curving gutter as a kind of a plastic constructivist idea, we created this landscape through this notion of art. It was a different way of thinking about infrastructure. That the sidewalks would somehow rise up and water would go under.

The county was about to spend millions of dollars on street infrastructure. They were putting in this big pipe to take away all of the stormwater. We started then having meetings with the engineers to basically buttress that work with this new work. We had all this momentum moving and then the political downfall happened in Opa-locka. There was a lot of corruption in the government and … everything stopped. We came up with this paint idea because we were feeling like the community didn't feel like they could do anything big. We're like, "Look, we have one day. We can go out here and we could paint a one mile stretch of road. And if we can do that in one day, think about what you can do if you had infrastructure money." **The paint project on one hand was a critique of the inequities of this kind of public realm, but it also tried to show them what they could do as a community.** And that was the first time we had ever worked in this way through art as an advocacy approach.

Ali Baba Avenue in Opa-locka during the paint day project led by the OLCDC (Ten North) and Hood Design Studio. Credit: Hood Design Studio.

Part of our work in a lot of Brown communities is to try to find ways to get people to see the inequities that they're normally just used to. People were not even thinking, "Oh, we don't have street trees," because they never had it. Where you go to upper income areas, it's normal to have street trees. Right? It's like the frog in the boiling water. You just get used to it. And before you know it, you're dead. And that's what's happening in a lot of these communities because these have become normalized conditions. So this idea was, *can design be a way to get people to see their place?* And through painting, it reveals something. It's this notion that if you put something in a place people will actually see the place differently. We were hoping that if things had gone on, then maybe through that project, they would have buttressed that infrastructure money with the street landscape and we could have had this amazing product.

. . .

Florida is paradise, you know, and it's people's imaginations. Very early in the 20th century, someone wanted to create Babylon. And so Opa-locka was that. A few of the buildings remained. If you thought of the character of the entire place, that was really all they had as an *identifiable* trait, and so the CDC saw that as something they could use. Which I think was the right way to go, because in places where identity is fledgling you got to hold on to something. And what was interesting about them grabbing onto the Moorish idea: You could talk about Blackness, you could talk about North Africa, you could talk about this kind of cultural diaspora right where it can inculcate the community and give them a sense of pride, a sense of dignity, not through our normative channels.

We were interested in telling this larger story about Brown people in this country and how design isn't just a mural on the wall—**design could be something that's tied to our culture and our diaspora**. This is something that

I truly believe is missing from landscape architecture, even our history. We've missed that whole North African diaspora and how the Moors went to Spain and actually created these aesthetics. How you had Brown people and Jewish people working together, creating these artful things. And all of this is left out of our conversation about design and the diaspora of colonialism, how it can be a place where you can actually grab and pull things out. That was something that we were really interested in.

And that's why, even when we painted, we wanted to double back down on pattern and color. Like being in a Moorish garden where light and things play out. We were just trying to use the culture of landscape—something that we know—to actually begin to elevate the project.

People romanticize the Jim Crow era in Black and Brown communities. They go, "Oh, that was a club here, there was this here." That's all people had, so that becomes the gilded moment. But how do you unpack all that? No one really wants to have that conversation. We're working again in Jacksonville, where the Black community was completely erased. And the only thing left is the club, right. The club. That was the theatre. But there's a romanticism about it and we're like, "Well, what happened to everything else?" We're making a Heritage Trail, trying to talk about those other things, but there's nothing really there. It gets to a point where it's just really impossible to do anything because you can't get people to have that vision without having this larger conversation.

If you don't deal with these issues of race and culture in this country, we can't get past the transformation of these neighborhoods. Particularly what we're seeing now with COVID-19, we know that early American cities in post-World War II, and even before, had red-lined neighborhoods that are *still* not rising and being invested in. And until we face up to that you can do all the CDC and planning you want to; it's just going to fail. For fifty-plus years there has been disinvestment and the placement of the most offensive and debilitating infrastructure that has caused so much trauma to people.

Dr. Willie Logan, President and CEO, Ten North Group
(Called Opa-locka Community Development Corporation until 2023)

I think to me, the most important thing for community-based organizations to focus on is their financial health. They need to develop a strong balance sheet and to develop an income statement that has earned, unrestricted income. You can't help anybody else if you cannot first take care of yourself.

In 2007, in the midst of the economic downturn, along with our organizational partners, we suddenly got the third largest RFP grant in the country behind the states of California and Michigan. The RFP grant we received was larger than those received by the county, larger than the city received, larger than the state received. We brought in some of the best consultants, including people like Reese Fayde, to help us make a really thorough assessment of what the people we served thought they needed and what the data said that they needed. And, what opportunity there was to meet those needs and who could actually help us accomplish that. Through the process, we discovered that we had no quality early childhood education centers in the community, and that the school system was putting in less money and less resources into our community, and the kids obviously weren't doing as well. We also recognized that the large absentee rates were due to the fact that these kids lived nine or ten miles away from the school and the School Board would only provide buses if the kids lived no more than a mile away from the school. So essentially, the kids were crossing major roads with trucks and stuff and they had to walk on streets that didn't have sidewalks or good drainage treatment when it rains, and this was what was causing them not to be able to go to school. Think about it. In South Florida, it rains from May to September.

Also, we uncovered that the health statistics were horrible. People thought they were healthy, but everybody living in a particular census tract had asthma. They would have skin diseases, hypertension, obesity. We uncovered all these wellness statistics that were bad. So, we had to step back, and say oh my goodness, that's a lot. Where do we start? **We realized that we could tackle the symptoms of an unwell community—which is something that many, many organizations do—or, we could go out on a limb and tackle the root of the problem**. So, we set about to build the prosperous, livable community we wanted to see, from the ground up. We started bringing in artists and infrastructure architects and public place-making folks between 2013 and 2016 to help us to visualize, through a very community-oriented exercise, strategies to meet the needs of the residents. But we also focused on things that those of us who understand data and statistics knew that the residents *needed*.

The architect and the urban planners and artists came and talked about physical improvements and events or activities. The residents said they wanted a place where they could go out, so we created a place where mommies and daddies can go on date nights and children can go and friends can go. We bought an old house and converted it. But the community said that the more important issue for them was education, access to healthcare, emergency resources, before we think about entertainment, culture, art—the infrastructure. So, we did what we call intermediary strategies and long-term strategies.

The services we provide are expensive. We provide wrap-around social services. And to do so, we have a staff of about nine or ten very dedicated and experienced professionals. And you know, wrap-around really means wrap-around. We do triage, like giving someone emergency assistance. We pay an electric bill. We pay rent. But that's not where it stops, nor is that what we want our clients to come to us for. We provide emergency assistance, yes. But we demand accountability. We demand that our clients do a deeper dive to get to the core of what got them in that situation. We provide counselling because somebody has a substance abuse problem or whatever the case may be, and then, once we go into the house, this is where the most impactful work really starts. We deal with everybody in the house. And because you can't fix one problem without addressing the household issues in general, and one is usually piggybacked on or impacted by another, we look to tackle the entire picture. We provide daycare services and transportation. Wrap-around services means we have become a full service, holistic organization.

We do a bunch of financial literacy. We teach our clients about budgeting. If somebody tells us they have a little side gig or an entrepreneurial venture, we try to provide them with technical assistance and a small grant. We may also be able to provide a small business loan, which can be vital to turn the side gig into a full-time job, or take a small business to the next level.

. . .

We will no longer be called Opa-Locka CDC. Our new name is Ten North Group. Our company is still the same because we have built up a lot of good will over forty-plus years. But, we needed a new name to reflect that we have evolved as a company and that we serve a larger community. Most of the real estate we own is outside of Opa-locka. Most of the people we serve are outside of Opa-locka. Also, historically, the concept of "Opa-locka" has not always been clearly defined. Outside of the City of Opa-locka, there was a wider [unincorporated] area that now is south Monument City, Miami Gardens. They had the Opa-locka post office and received the Opa-locka services so they thought they were from Opa-locka, but they couldn't vote. It was a community more so than a city. Secondly, we, the City of Opa-locka, had very difficult times for any number of reasons over the past twenty years. Those difficulties have been conveyed in the press very negatively, which resulted in pre-conceptions that we needed to overcome (unnecessarily). We also recognized nobody knows what a CDC is. And just as importantly, we believe that we have evolved beyond the confines of a CDC—with a model that I am proud to say is relatively unique and at the cutting edge.

The reality of it is we are much larger than Opa-locka and, with our new name, we can honor our history—since Opa-locka is located ten miles north of the City of Miami—while at the same time, not having to justify or explain before we talk about what we want to do. The work that we do deserves to speak for itself, to tell its own story in its own way and not be saddled with any of the noise. And Ten North Group, because it can be anything, is definitely a conversation starter.

. . .

We've accumulated a significant enough amount of property in Downtown Opa-locka—including property across from the railroad station and we have enough property now for about 500–600 units and about 100,000 square feet of commercial space. We also have planned another 50,000 or so, 25,000 or so, square feet of community space. A real concern of ours is gentrification: Opa-locka has the only urban tri-rail station in a fifty-mile radius and now they are building a direct line from Opa-locka station to Downtown Miami. Once that happens, everyone is going to want to come here. And so, part of what we want to make sure of is that Opa-locka doesn't become another town like Wynwood or Coconut Grove—where gentrification meant displacement, since the price for revitalization was that the long-term residents were overwhelmingly pushed out, as there was no game plan in place to make sure that they could participate in the new prosperity. So, we want to avoid repeating the mistakes of all these little communities in Miami-Dade County which have now become gentrified in a way that we don't think is good for the long-term health of the South Florida Region. It's not healthy for the region if gentrification continues to result in neighborhoods for the wealthy.

. . .

The object is to get 100 percent ownership if we have apartments. This is because if **we own these assets outright, we can control their destiny. We can ensure that they continue to be affordable housing**. The way to achieve this 100 percent ownership is by making sure that even if we develop apartments with a partner, we have the legal right to buy that partner out, and to do so at a reasonable price (not market price, but a reasonable price). That only happens if we negotiate a right of first refusal from the start. Back in the 1990s, we did not have the leverage to demand the right of first refusal to purchase the real estate projects we were involved in, from our partners. In one case, Sun Trust were the majority owners in a deal where we did not have the right of first refusal. So when it came time to sell, we could not stop the sale and they said they would be converting the project to market housing. They basically said: be glad and take the windfall, and that we should get twice the value even though we lost 300 or some 400 units of affordable housing. We would have never done that, if we controlled the destiny of the project. We would have kept the project affordable even if it meant less money in our organization's coffers. We don't have dividends we've got to pay to shareholders, so we're not motivated to cash out at the highest price and sacrifice the most needed use for these properties—which is to keep them affordable.

You go to our website and it's mostly all about our services—the business empowerment, arts and culture, financial empowerment, and family services. We don't showcase the real estate deals or our balance statement. But we spend a lot of time in meetings talking about real estate because it's the real estate that makes our mission sustainable. It's the real estate that facilitates the business empowerment, arts and culture, financial empowerment, and family services. Every real estate contract provides resources to our residents, to our school programs, to the city kids we work with. They all benefit directly and indirectly from our real estate funds. And so, I think everybody's beginning to understand what we have known for a long time. They are beginning to understand just how very important it is to have a

Access to downtown Miami, Florida, via Tri-Rail at the Opa-locka station.

3.2 Opa-locka, FL

sustainable financial model. Without a sustainable financial model, no matter how good the strategic plan may be, you can't carry out good ideas and visions.

Reese Fayde, Project Consultant for the Opa-locka Community Development Corporation Federal Grants

We started with all bets on the table. What could this community look like, what could they use as their organizing approach? And in that conversation, we moved toward the arts, in part because we had these marvelous buildings all around. One of the evenings we were there, we said to the director (Willie Logan), "Wouldn't it be great to go over to Miami and take a look at this art fair kind of thing." Their version of Art Basel. And he went "Ahhh, not my cup of tea." And we said, "C'mon, nationally this is a really big deal." So we dragged the poor guy over. And it was literally mind-boggling. It's everywhere and it's everything. And you could sense the power of having an art show to be able to draw literally thousands and thousands of people. And after that, the director agreed about having arts be part of his revitalization. He has since become an absolute art collector, and a real aficionado. To the point that, if you go onto their website, they have an art section. They have a curator. They have taken it on very seriously. And one of the people who participated in the charrette with us had an art gallery and an art collection, which he has since donated to the CDC, so they now have an African arts collection and a space.

. . .

Opa-locka's needs were so incredibly basic: employment, employment, employment. You have growing industries around you, but your residents are not part of that success story. Within Opa-locka, there were a lot of repair shops. If you had a motorboat or an airplane part, you would go to Opa-locka to find a mechanic. But the streets weren't paved, there were huge drainage problems. The housing was incredibly modest. Not something which would be picked up by the artist community as you know, artful and "Gee, wouldn't

"Beyond Tradition," a show at the ARC in Opa-locka. This venue, among others, is supported by the CDC in their efforts to center African and African American designers and artists in every facet of rebuilding.

we love to go here and help." I don't think they were managing gentrification—I don't think that was part of their immediate problem.

. . .

Opa-locka's downtown has at its edge a train station and a train that runs by all the time. You'll be talking and have to stop because the train is going by. But this is a train that takes you into downtown Miami and to other parts. And I remember any number of times talking about access to jobs and being able to take the train. And everyone looked at me like I was crazy. Because no one that I ran into ever used the train. They are car-oriented. We'd say, "But you can go to the main Miami airport right here! This is so great!" and it was like, "Really? Nah, I just drive there." I believe Willie was able to open up a childcare facility in the train station, which was great idea, but I'm not sure the train station was being used for transportation as much as it was for other things.

Danielle Levine Cava, Miami Dade County Mayor
(Elected November 2020) with Sean McCrackine and
Jorge Damian de la Paz, Mayor's Office Staff Members

Miami Dade County government serves almost three million people and thirty-four cities. Opa-locka is one of the ones that has been struggling and also has very low socioeconomic households, but it has unique charm and given the huge growth of the county overall is poised to really have a recovery. There are a few factors that lead to that. One is certainly just that it is in Miami Dade County. Another is that the land is not as expensive as in other places. It does have a commercial base; we are dependent on property taxes in Florida, no income tax, so we get a portion back of sales tax dollars that go to the state. But primarily we are dependent for local government on property taxes so there is some base there, but the residents have very, very low income, and building back has been complicated. The city has had numerous political challenges. But along with this political upheaval and frequent turnover of elected officials, there's this mismanagement of essential resources like the water and sewer system.

We do have a number of county departments that provide services in Opa-locka so we have a significant social service presence. Like our public housing dollars and oversight for some of them, or partnerships on some of the affordable and workforce housing projects. Also, our transportation department provides bus service and a couple of different transit corridors which are also very important for Opa-locka's growth. They have the Tri-Rail train, which is a commuter train with connection to the buses to get downtown and elsewhere so it is an important transit commuting route. And we are looking at other ways to enhance that as more people take the train. And then we will be, God-willing, knock-on-wood here, building out a North corridor extension of our metro rail line and there will be a stop that would service Opa-locka. So, there's definitely opportunities to improve mobility, workforce housing you know that could rely less on the car and there are also jobs, right, so it is a job center as well.

A CRA [Community Redevelopment Area] is basically taking general property tax dollars from the area as it grows and allowing a percentage of it to be kept within a district that has been identified as suffering from slum and blight for the idea of doing focused redevelopment in that area. So, there is an Opa-locka CRA, an agency which will be in a position to invest dollars in infrastructure in Opa-locka. They're supposed to have a plan and they have a budget approved by the county. So, we do have some oversight which has been kind of lax in the past, but I have added expertise in my administration so that these CRAs would actually be productive, because in the past sometimes the monies haven't yielded much for some of the CRAs. I believe in them [CRAs] because I do think that targeting investment is helpful rather than just coming into the general fund and having to fight for special projects within the county budget.

. . .

All of Miami Dade County is vulnerable to gentrification and it is a really big problem because we don't have strong laws protecting neighborhood preservation. So I offered changes to our CRA legislation to try to offset some of the gentrification pressures: local hiring, set asides for low- and moderate-income workforce housing. Those things did not previously exist, they maybe assumed you were going to take this money and do something that would be inclusive but it was not a mandate. So we did build it in for new CDCs to have more controls that would fight some of those trends. I think the fact that Opa-locka's CDC was in a way the first to discover Opa-locka is a good hedge, because no matter what comes next, you will have some concentration of more affordable housing.

. . .

The county government is changing, it's not a consolidated government. I don't have powers to override cities on their local land use and zoning, for example. But we use the same transit, healthcare, major parks. We are in charge of environmental issues for the whole county, the airport, the sea port, a healthcare system, the libraries—there are many, many general things. We serve fire [protection] for most all of the county and police at least for the majority of the county. So, there's a lot of overlap in those regulatory and contractual functions. It's really critical that we work collaboratively, so I consider myself collaborator in chief.

There is a *League of Cities* for the county and all of the cities participate. I am now an active member in that. We have a very active intergovernmental team for constant communication. We work together around COVID-19, around crime, around water, stormwater issues. I'm working hard to try to get all the cities to collectively work on stormwater because each city is actually responsible for their own stormwater drainage. There are some sister city relationships. Miami Gardens, which is a very successful large African American city in the state of Florida and a newer city, is paired with a richer city that's more suburban in the southern part of the county. There's a best practices conference. I've worked very hard to be a part of the planning team.

References

Barnes, G. (2020, September 8). P. Heyda and D. Gamble, interviewers.

Barnes, G. (2021). *STUDIO Barnes*. Retrieved from https://germanebarnes.com/Made-in-Opa-locka

Carnick, A. (2022, December 21). To Miami with Love. *Design/Miami*. Retrieved from https://shop.designmiami.com/blogs/news/to-miami-with-love

The City of Opa-locka Florida (n.d.). The City of Opa-locka History. Retrieved from www.opalockafl.gov/229/History

Davis, C. (2020). P. Heyda, interviewer.

Florida, R. (2002). *The Rise of the Creative Class*. New York: Basic Books.

Florida, R. and Pedigo, S. (2019). *Toward A More Inclusive Region: Inequality and Poverty in Greater Miami*. Miami: Miami Urban Future Initiative, Florida International University.

OLCDC (2019, September 5). Transforming Magnolia North Park. Retrieved from www.olcdc.org/post/transforming-magnolia-north-park

Pacheco, A. (2017, April 14). Fate of HUD Imperils Revitalization of an Iconic Miami Neighborhood. *The Architect's Newspaper*. Retrieved from www.archpaper.com/2017/04/hud-made-in-opa-locka-neighborhood/

Ten North Group (2023). Creating Stronger Communities Since 1980. Retrieved from www.tennorthgroup.com/about

United States Census Bureau (n.d.). US Census Bureau Quick Facts. Retrieved from www.census.gov/quickfacts/fact/table/opalockacityflorida/HSG495219

VenusInOrbit (2016, January 31). #OpaLockaArt: How Art and Culture Positively Transformed One of the Most Dangerous Cities in America. Retrieved from https://venusorbitsforjustice.tumblr.com/post/138454147749/opalockaart-how-art-and-culture-positively

World Population Review (2023). Opa-locka, Florida Population. Retrieved from https://worldpopulationreview.com/us-cities/opa--locka-fl-population

Section 4 **Outposts**

4.1

Traverse City, MI

In Michigan, 50,000 people or less is the benchmark for a *micropolitan* region. Traverse City has a population of 15,600. It rings the shores of Lake Michigan, one of the five Great Lakes that hold 20 percent of the world's supply of fresh water and provide sustenance, work and recreation for tens of millions of Americans (Egan, 2017). As the largest population center in this forested, northwestern stretch of the state, Traverse City's temperate climate, verdant forests and natural aquifers provide an ideal climate for vineyards and fruit orchards. The region is the number one producer of cherries in the United States, and Michigan is the second most diverse agricultural state in the country. Nicknamed the Cherry Capital of the World, Traverse City's week-long National Cherry Festival draws close to 500,000 people every summer. The city is a local treasure, but also a desirable tourist destination nationally.

In the middle of the 19th century, Traverse City's relationship to the natural landscape made it an ideal location for an innovative new model of hospital design for individuals with mental health needs. The Traverse City State Hospital opened in 1885 and served a large part of Michigan for over a century until it closed in 1989 and was gifted to the State of Michigan. At its peak, the sprawling campus of gabled stone buildings had a population of over 3,000 residents—nearly half the town's population at that time—and it was the largest employer in Traverse City.[1] The complex resided on hundreds of acres of parkland less than a mile outside of downtown with creeks and aquifers and a working farm for the patients and staff.

During the 20th century, the large institutional model of housing mental health patients became increasingly obsolete. Many were overcrowded, and patients were redistributed into smaller facilities. That era also saw an increased reliance on medication for treatment. By the late 1970s, deinstitutionalization caused the closing of such mental health facilities around the country and Traverse City's immense, 800,000 square foot complex and its associated cottage buildings fell into disrepair (Dowdall, 1996).

View of Traverse City looking out over the Grand Traverse Bay of Lake Michigan.
Credit: Stone Hut Studios.

Building 50, the centerpiece of The Village at Grand Traverse Commons, a mixed-use community retrofitted into the footprints of a former state mental health hospital in Traverse City, MI.

Saving the complex was a public–private effort. In 1963, the ornate and iconic central Administration Tower of what was called Building 50 had been demolished and replaced by a low-slung one-story addition. Over the next three decades, the vast hospital campus was threatened but saved several times over by various concerned citizen groups who crossed public–private sectors, including Traverse City's "community caretaker," preservation activist and former Mayor Carol Hale, who was said to have stood in front of bulldozers to stop the hospital's demolition (Siciliano, 2021). Such advocacy efforts were part of getting the hospital listed on the National Register of Historic Places. By 1998, the Grand Traverse Commons Development Corporation (GTCRC) recommended demolition again, citing its redevelopment as unfeasible. However, a "Committee to Preserve Building 50" (CTPB50) was formed to advocate for its preservation. That fight was protracted, and a group of local architects and engineers emerged with ideas that showed the feasibility of rehabilitating the massive complex. As a result of the group's advocacy, demolition was averted.

It would take another few years before the GTCRC would vote to convey the complex to the Traverse City-based private developers called the Minervini Group who promised to preserve and redevelop it. In 2002, portions of the campus received new roofs to avert additional deterioration and redevelopment of the complex began in earnest. In 2019, Ray Minervini, President of the Minervini Group, was thanked for his company's commitment to rebuilding the former institutional facility and he was awarded the key to the city from Mayor Jim Carruthers (The City of Traverse City, 2019).

In 2024, after more than two decades of continual investment, the "Village at Grand Traverse Commons" is a massive residential, office, fine dining, commercial and retail hub within the 19th century former asylum. The backdrop of the campus is a large wooded area, part of the 480 acres of reserved parkland that includes a formal lawn in the front of the campus. Downtown Traverse City with views of Lake Michigan is a relatively short walk away. The complex is just far enough away with enough critical mass of uses to establish an autonomous identity without competing with existing downtown retail. The park-like setting is attractive to visitors, residents and office tenants alike. Like its name, the complex functions as a village within an expanding town nested within a highly desirable region.

The Minervini Group created space for a wide range of functions and income levels within the shell of the former institution. The residential portion deliberately spans economic brackets; there are luxury, market rate, affordable, senior and low-income units. Many of the former outbuildings beyond the main building have been adaptively reused into coffee shops, bakeries and wine bars. And the grounds—which were once farmlands related to the asylum—are attractive gardens for passive recreation and seasonal events. However, the reoccupation of the complex was far from easy. **The Traverse City case demonstrates the challenges and benefits of adaptively reusing large, historic buildings for contemporary needs within the context of a small town.**

Landscape Connections as Therapy

Thomas Story Kirkbride, the founder of the American Psychiatric Association and chief physician of the Pennsylvania Hospital for the Insane, was a visionary and influential figure in 19th century asylum care. For more than forty years, he headed what was considered one of the finest mental hospitals in the country. His most enduring legacy to the proper care of the mentally ill was the development of the "Kirkbride Plan," a style of hospital design and management that shaped the first wave of asylum construction. The method became a model that was deployed nationwide in the mid to late 19th century and forms the basis of the plans at Grand Traverse Commons as well.

The primary composition of the hospital was laid out in an east–west alignment, centered on a main administration tower and chapel in the middle[2] Characteristics of the design typology included ample southern exposure for patient treatment areas, including large hallways that were used for therapy. Individual buildings, called wards, were separated for fire safety. Natural ventilation was prioritized and integration of the building into the surrounding landscape was created through the incorporation of working farms providing recreational and therapeutic opportunities for the patients.

The Old Barns, once part of the hospital's farm area a short walk away from the main complex, has been renovated for special events together with a new Botanical Garden pavilion.

As of 2024 there are approximately thirty remaining psychiatric centers of this kind from the more than seventy that were constructed from 1860 to 1900.[3] The complexes, like other large abandoned buildings across the country, increasingly face demolition in light of their considerable size, levels of deterioration and the general challenges associated with adaptively reusing functionally specific buildings for contemporary demands. Some of them, like in Athens Ohio, have been occupied for educational use (Ohio University). Others, like in Buffalo, New York have brought in a boutique hotel, architectural center and conference venues to anchor the central part of the complex. The Kirkbride in Weston, West Virginia (formerly known as the Trans-Allegheny Lunatic Asylum) today offers seasonal "ghost hunts."

Starting Small

When Ray Minervini and his development group sought to save the 500,000 square foot abandoned complex from demolition in the early 2000s, it had already deteriorated from decades of neglect. In addition to re-roofing the entire complex, the highly visible, outermost 45,000 square foot building was then fully restored, as this portion of the complex was one of the outlying wings and highly visible from the ring road that defined it. The ward structure formed the cornerstone of the redevelopment as it demonstrated an adaptive reuse strategy that led to future phases and propagated the initial vision. Minervini's strategy for redevelopment of a complex this size was to tackle it incrementally, building in phases to reinvest initial returns into the stabilization and future phases of transformation. Some portions of the complex were sold off over time to provide working capital renovations elsewhere, although the bulk of the complex remains in the developer's control. There was no formal "master plan" for the Village at Grand Traverse Commons.

A room in the old hospital before renovation. Obsolescence, scale and severe deterioration were some of the challenges in restoring the former mental institution. Credit: The Minervini Group.

Towns that have such historic institutional buildings in a state of disrepair face significant hurdles to redevelop them. The cost of rehabilitation is considerable given the buildings' age, and many institutions like this are well over a century old. The massive scale of the buildings—compounded by their specialized and idiosyncratic layouts—make adaptive reuse difficult. If a developer seeks subsidies in the form of state or federal historic tax credits, which are almost always necessary to make the rehabilitation economically viable, they must follow strict guidelines on the type of interior or exterior modifications that can be made. In addition, many 19th century Kirkbride hospitals were built on the outskirts of town, where today there may be limited evidence of market demand or where economic realities can make attracting viable tenants problematic.

Numerous economic incentives helped offset the high price of renovations, and the capital stack was multi-layered. The project was identified as a Michigan Tax Free Renaissance Zone in which residents and businesses paid no state or local personal property or income tax for fifteen years. The developer qualified for state and federal historic tax credits for rehabilitation, which in Michigan were very favorable for decades until the program was ended in 2012. (A new state program was initiated again in 2020.) Brownfield funds were also directed to Traverse City from the state capitol in Lansing.

Smaller scaled cottage buildings outside of the main complex were redeveloped early on as an easier economic lift, and to bring life to the areas while other renovations continued.

From Hospital to Hospitality

In the contemporary American political economy, mixed-use projects collectively create new market dynamics for themselves but are arguably only successful when they serve as stewards to their host neighborhoods—in meaningful social, qualitative and environmental ways.

Although physically separated from downtown Traverse City, the Village complex is adjacent to the area's primary medical center, Munson Medical Center, a modern health care complex that employs thousands and enables synergies with the village campus. The mixed-use Village continues to grow incrementally over time as momentum builds and as resources become available. As of 2024, there were numerous restaurants, a local brewery and winery, a secondary school, office space and a wide variety of residential units. A quarter-mile-long retail "Mercato" in the building's former basement level

The basement level of the historic hospital complex had corridors to connect the various ward buildings. The quaint brick passages were restored into an underground "Mercato" for small businesses and social spaces.

unexpectedly seeds a continuous line of commercial activity, linking each ward building with galleries, salons, local maker spaces and other amenities. Future plans include the construction of a boutique hotel in two remaining vacant buildings with a restaurant and event space. It is a dynamic, evolving place where residents mix freely with tourists. The mix brings diversity, but also some tensions.

Most people arrive via car, and parking is at a premium. Much of the rear portion of the complex is impervious pavement to accommodate this demand. Ray Minervini underscores the conflict between vehicular needs and the more ecologically responsible, pedestrian-oriented environment he and his company have sought to create:

> Parking is a huge frustration of mine, because I didn't want to put as much parking in as I did. I wanted to plant more trees. The other thing we battle is that all the roads here are private. The maintenance of the roads, the calming of the roads, the streetlights, the raking of the weeds and the maintenance of the grass, they fall on the developer.
> (Minervini, 2019)

To address the auto-dependency of the site, in 2017 the public Bay Area Transportation Authority created a free bus circulator between the Village and the downtown waterfront (and other sites) that some Village employees and visitors use. Some laud the public provision of transportation, something not all smaller towns have—that also helps reduce emissions by reducing car trips. Others question the use of public funds for an alignment that concentrates along the waterfront and serves touristic destinations.

Love the Land

New growth on town peripheries can come at the expense of development in the downtown core. Greenfield development—which is attractive to investors seeking fewer encumbrances—is enabled by larger land parcels and often lower land costs with easier access to primary transportation corridors. Such features simplify construction sequencing and streamline permitting processes, even as they require the need to extend utilities like water, sewer, electricity and street infrastructure. Most importantly, however, development on the edges erodes the natural landscape—and the critical ecosystem services it provides more broadly. And even when municipal edges are not developed with new buildings, they can still become commercialized by large-scale agriculture that creates other kinds of environmental burdens.

The Grand Traverse Land Conservancy in Michigan has been a vital player in preserving large swaths of land through land trusts that hold, in perpetuity, the naturalized landscape and protect it from low-density, suburban sprawl. The non-profit organization—whose motto is "Love the Land, Pass it On"—has raised over $180,000,000 to help preserve, with its community partners, over 40,000 acres and 121 miles of shoreline in the Grand Traverse Regional Land Conservancy Service Area, of which the Village at Grand

The sprawling campus of the former mental health institution, now a mixed-use community in Traverse City, Michigan. The town center and Grand Traverse Bay are less than two miles (3.2 km) away, seen in the distance and to the left of the frame.
Credit: The Minervini Group.

Traverse City is a part. Fernando Meza is a business owner in the Village. He describes the importance of the regional landscape in even more urgent terms:

> I appreciate what the Conservancy is doing, getting these big, big swaths of land and protecting them with living trusts. Losing them would be such a shame because we are sitting on, what is it, 20 percent of the world's freshwater? We have these mass migrations all over the world and conflict because of climate change. When people can't grow food in their areas they're going to go to where they can. It's that simple. Fifty or sixty years from now, what is this place going to look like? There is a potential for all this (landscape) to be compromised.
>
> (Meza, 2019)

Land conservation in small towns creates exponential benefits, helping both the local ecology and the local farm systems thrive. Even as Traverse City undergoes pressures to accommodate more tourists, it seeks to protect its resources for future generations.

Rebuilding Is Carbon Capture

It goes without saying that building reuse is a critically important design strategy for combatting the current and future climate crises. Reuse provides up to 46 percent greater environmental benefits than new construction for comparable, even "green" new buildings (Preservation Green Lab, 2011). Reusing an existing building preserves embodied carbon and limits demolition waste and carbon-intensive new material acquisition and transport. Embodied carbon saved is carbon emissions avoided (Siple, n.d.). The reuse of such an immense complex at the Village at Grand Traverse, with its new programs and quantity of residential units, makes it a particularly viable model with exponential sustainable impact.

Locally Isolated but Nationally Connected

As an "outpost town" Traverse City is a four-hour drive from Detroit, Michigan's largest metro area. Traveling northwest through the state, one passes through extensive forested areas. However, Traverse City is not completely isolated. To the surprise of many, there are direct seasonal flights to Traverse City from eleven national airports, including: Newark, LaGuardia, Washington, D.C., Denver, Charlotte, Indianapolis, Dallas, Chicago, Atlanta, Philadelphia and Toronto. For better or for worse, for people of means who live in New York City, it has become easier and much more economical to fly direct from LaGuardia Airport to Traverse City, Michigan and buy a million-dollar lake-front home there, than it is to slog through a long traffic jam from Manhattan to Long Island, New York and pay five times as much for half the house. This picturesque small town becomes a big-city weekend escape, enabled by the airport—ironically a major contributing industry that threatens the environment but collapses distances and time. **During the COVID-19 pandemic, small towns like Traverse City became more permanent getaways, where many home-bound executives conducted global work from second homes with a view.**

As alluring as The Village at Grand Traverse Commons is, and as vibrant as the downtown can be in the height of the summer season, a tension exists between those who have always known it as home, or even as a regional vacation destination, and the new influx of people that are perceived as

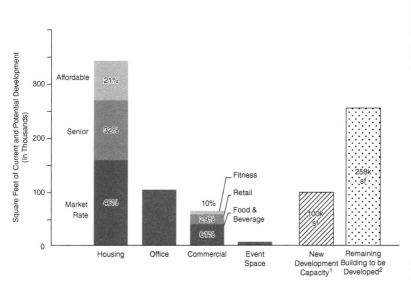

The Village at Grand Traverse Commons
The former asylum distributes residential, office, retail and revenue-generating event spaces throughout the 816,000sf complex. *Source: Minervini Group*

[1] Estimate, based on local development zones and State Historic Preservation Office approval.
[2] Under roof, not yet placed into service.

changing its character making it more generically hip and urbane. In effect, the town is becoming unaffordable. Old time locals dub it "Tragic City."

The city's attractiveness is a double-edged sword. At one time, Traverse City was less picturesque. Glen Chown, Executive Director of the Grand Traverse Regional Land Conservancy explains:

> This city really had some important choices to make during the 1960s and 70s. Traverse City was wall to wall industrial buildings and the West Bay was red because of the cherry canning plants and the lumber mills and the power plants. We were as industrial a Midwest City as anyone. It was a conscious choice to move all of that industry off the waterfront and to create business parks inland, a couple of miles away from the waterfront, and open up the waterfront. If you didn't do that, none of this, I think, would have happened.
>
> (Chown, 2019)

The town has grown 7 percent from 2020 to 2024. As new resident profiles grow and change, Traverse City grapples, like other American towns, with how to build—rebuild—equitably and for everyone.

Perspectives

Glen Chown, Executive Director, Grand Traverse
Regional Land Conservancy

Opening up the waterfront made Grand Traverse Bay our front door, our best asset. And then, Rotary Charities bought the Park Place Hotel and invested millions of dollars into fixing that up as an anchor to downtown. Horizon books bought the J.C. Penney's store and turned it into a bookstore/cafe. Several other important renovations and purchases in downtown Traverse City stabilized the whole urban core. And then Ray Minervini showed up with this vision for the former asylum. But it was not until there was a group of civic leaders that formed the Redevelopment Corporation and literally stood in front of the bulldozers when they were going to raze this complex and said, "You're not going to do this." They started negotiating with the municipalities and then the state of Michigan. There's a very active historic preservation ethic here.

    Alongside this project, we're protecting lands around the urban core, particularly the farmland. The goal is to preserve 9,400 acres of farmland. The peninsula itself is 16,000 acres or so. And we're now at 5,700 acres preserved, approaching the 60 percent threshold. We have some of the most unique growing conditions, and various fruit crops: cherries, peaches, apples and grapes. And, unlike California, we have water. Look at those Bays, they're carved by glaciers—in places they're 600 feet deep. The water quality is still superb, and there are no industrial intake pipes anywhere in the 132-mile shoreline. We're sitting on 20 percent of the surface water on planet Earth. So you preserve this and create a whole new life, not just the value of the crops, but agritourism.

    People need to live in the core, where the infrastructure and services are. It's a European kind of model. But now our biggest challenge within Traverse City is affordable housing. The real estate values are just going through the roof. We're gonna have growth pressures. We've got transportation issues. We've got these two bays that bottleneck and there's no bridge over them, so you've got to go through Traverse City. Traffic is becoming more of a problem.

    Our staff is about twenty-seven now and our operating budget's a little over three million dollars. But the last three years we've been raising over $10 million a year. We have thousands of donors. We have a really solid major gifts program and foundations have supported us. We have a great estate planning program for planned gifts and requests. And we've also been able to get a lot of public money for land transactions through the Mid-Michigan Natural Resources Trust Fund, which is generated by oil and gas revenues on state land. I've known some of these large philanthropists for twenty-five or more years, and they feel like they're part of the conservancy family. There's a high degree of trust. You're not going to make a really big gift unless you *trust* that the organization is going to steward that gift well.

    Because we're a charity, we can buy land at a discount, so the landowner can take a donation for the discounted value. We only work with *willing*

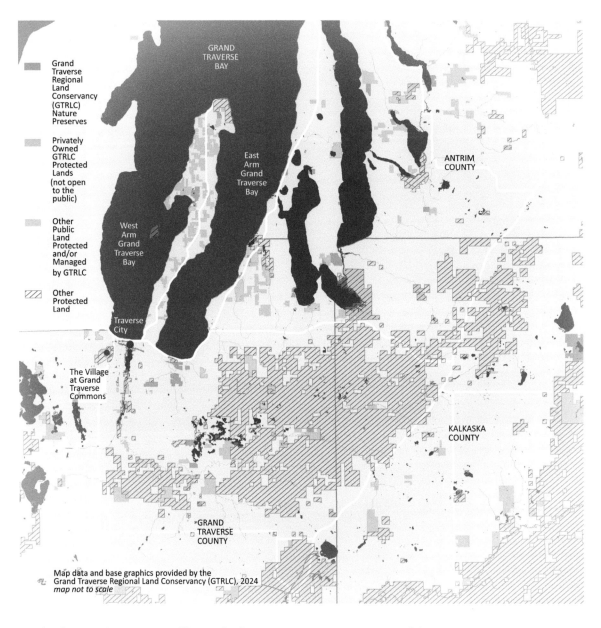

landowners. Sometimes we'll have talked to someone twenty years ago and then all of a sudden they'll call and say, "Well, now we're ready." Maybe they're not doing well and they're trying to get their affairs in order. So, our work is focused on relationships: with our donors, with our communities, with landowners.

Christine Krzyston, Resident at The Village at Grand Traverse Commons (Travel Writer)

I wanted to live where I was close to things to do, and I wanted to be around people. I didn't need the 2,500 square feet I was living in. And so, when I heard about this project, I thought it would be perfect to downsize. That was in 2007. The people that moved here originally really took chances because the economy went down and nobody was doing *anything*. You could have lost everything or you could have been living in a construction zone. Fortunately, the downturn was short. It was a real risk, but it turned out to be the best decision in my life.

The main complex is more traditional with fourteen-foot ceilings and nine-foot windows; it's very old. My place is contemporary, which is more my personality. I like change; I've been all over the world. I go on three international trips per month and I always think when I'm gone—"Could I live here?" But Traverse City is better.

My friend does consulting with the airport and you would not believe the status that Traverse City has in the airline world. There are eleven direct flights. We have Delta, American, United, Allegiant. For a town of 20,000 the level of air service is unprecedented. It's one of the reasons I'm able to do what I do.

It's actually very surprising to me how well it gels. You have subsidized housing, you have senior housing, you have renters, you have tourists, you have shops, you have businesses—you have every slice of a culture here. And yet, it truly is a village.

The nightly rental rate is high because we have so many events going on, so you can charge an astronomical amount. It's very lucrative. Some people live here in the winter and then move out in the summer and just rent it out all summer long.

We've all seen buildings where as soon as the people move out, it starts corroding and just wastes away. Here you're taking something and doubling its life.

Brick walls and large windows flood light into a condominium in a restored ward of the former institution. Residential units range in size from 300 to 3,000 square feet.

4.1 Traverse City, MI

Fernando Meza, Small Business Owner at The Village at Grand Traverse Commons (Digital Marketing)

The new Traverse City has some of the things that you look for in a bigger city like food and culture, and it's just a really dynamic area. A lot of the people that I'm hiring are young families. **Traverse City is this little speck of blue surrounded by a sea of red, with a really inclusive community.**

 I have never been as proud of my heritage as I have been since I moved here, and that's something that I did not expect. I mean, they don't call it the Great White North for nothing up here! But then you have these people who are like "Oh, you're from Mexico, that's amazing. I love that area!" They're genuinely interested in me and my culture and my community. That was a surprise.

 Granted, I was making over $100,000 a year living in New York and then I took a significant cut. I think my starting salary was $40,000 back in 2002. Surprisingly, the cost of living was so low that I was making less, but somehow I was getting out of debt faster. I couldn't believe it. Like you go out to eat here and you spend five or ten dollars and that's it. Whereas if you go for a hamburger and a beer in New York, good luck.

 . . .

 The community right now is very attractive to people that are retiring and they're driving up the cost of homes and condos and it's almost making it cost prohibitive for some of my younger families who can't afford to put 20 percent down on a $300,000 home. So the rent now is starting to increase.

 Whenever there's a developer that wants to come in and throw up a high rise that's taller than sixty feet, there are people that say "No, we have to preserve the character of Traverse City." Well, what is the character of Traverse City? Let's have that conversation. There is a lot of opposition to making this a more diversified community from a building perspective. You have these clashing cultures of people that grew up with what feels familiar and then these new, younger, more progressive people that are interested in pushing the envelope. That new demographic is going to have to figure out a way to make peace with the old guard of the community that is still, for all intents and purposes, in power. It's funny because even within the old guard there are card-carrying progressives, but when it comes to development they get very, very protectionist.

 We in the business community are trying really hard to make sure that Traverse City is not just a tourist destination. It's not sustainable that way. That's why we have Casey Cowell and TCB Tech NMC (Northwest Michigan College), a community college that has *amazing* programs. If you drive by the bay you see the Marina Academy that teaches people how to become captains of big ships.

Ray Minervini, President, Minervini Group, and Developer of The Village at Grand Traverse Commons

Mixed use was our goal from the beginning to make this as broad-based a social and economic area as we could, because typically the banks and investors like to see you're building 2,000 square foot units that should be all about the same price. But that just attracts people who can afford that specific range and you end up with a monochromatic community as opposed to something diverse. We had these little 400 square feet orphan spaces and thought, let's just make a little hotel room. We'll put a small sink, and a cooktop, a bathroom and a bed and that's it. We'll rent them out because nobody's going to want to buy them. Well, as soon as we built them they were rented and as soon as somebody moved out, boom, they were rented again. So, we knew we were onto something.

We had a little resistance from people here. So we said "You know Sally who works down at Stella? She's going to be living down there. And you know Bill, who works at the winery? He's going to be living over there." So they could put a face with a tenant. Until a short while ago there were more low income housing units here than market rate housing.

You know, this whole site had never generated a dollar of revenue for the city because it was owned by the state and it was a non-profit—the hospital, the school district, all non-profit. We're generating income so that's what we can take to the city and the community and show them "Hey, we're paying our way."

We don't make a lot of money on the Mercato on the lower level. But it adds such an element to the whole site. It creates a lot of traffic. You can tell prospective purchasers, "You can leave your robe on if you want to come down in slippers for a cup of coffee or to go Red Spire for a great breakfast, or you can go to the bakery." **It's all right here, womb to tomb.**

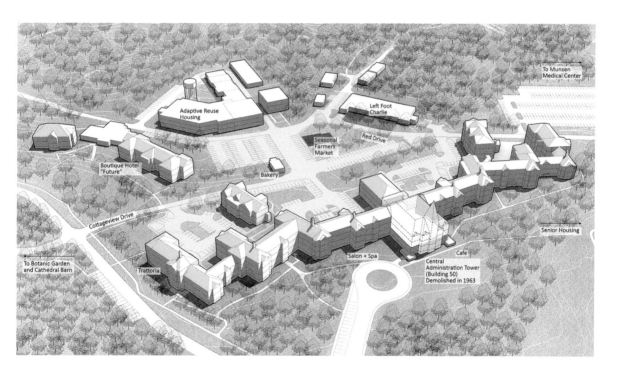

It almost doesn't pay to do this with the federal (historic tax) credit because it's such a pain in the ass. Especially when you're talking about a building like this one that's on the National Trust. They should be cutting some slack to these kinds of buildings. Do you want to preserve the building or not? Yes or no? Frequently, the answer that we got is "Well, we can't save them all."

First of all, there's got to be a discussion about whether or not the buildings are worth saving. And if they are, you need to create opportunities for that reconstruction to happen. There has to be some financial gain. Why do it and lose money? The success of the project is dependent upon its economic viability. So, our approach was one bite at a time, and mixed use. We wanted to be able to rebuild with retail on the lower level, commercial offices on the second floor, and residential above that—the way downtown Traverse City was built in the 1850s, when shop owners lived right there in the building. And they said "Well, you can't do that." And we said "Yeah you can, you just have to change the ordinance to allow for mixed use."

I knew that the reuse of the building had to be done and nobody was going to do it. I was on the committee trying to save it and I realized it wasn't going to happen. They had a developer who was going to pay a dollar for the property and in return was going to demolish all buildings and build some senior housing—vinyl sided shacks, basically. They had the votes and the approval. I went to them and said, I've got a plan you need to look at. I'll give you a dollar for the buildings, but that's a dollar to save them. Then I'm going to put forty-year shingles on the entire roof of the 500,000 square foot main building, and I'll provide a bond to ensure I do that. And then I'll start working on the development, and if I fail the building is preserved for forty years until you come up with another plan. How can you lose?

The community got behind that. It was a million and a quarter for the shingling of the whole roof, and that's the cash I had upfront. That was obviously not the most judicious deployment of assets with no return, but it was important for two reasons. First, we stopped the deterioration of the building, and second, we sent a signal to the community that it was going to happen. Why the hell would I invest all that money into a roof if I wasn't going to develop the *whole* building?

When I first started going around to the different banks they asked, how is financing going to work for the project itself? I said, we will condominiumize the building and take reservations for space and we'll have estimated costs for the build-out. So if you want a 1,000 square foot unit you know it to be X dollars and we'll collect a deposit that's refundable if we don't move forward with the project. Pretty soon we had half a million dollars. So I went to the bank and said "Can I deposit this here?" and when they said OK, I said, "Then I need you to help us finance this project." That opened the door. It's actually a small local bank, and the banker formed a consortium with three other small local banks and convinced them to participate as well. They did the first section, which was I think a three or four million dollar loan. And then the next section was like six, and in the next section I needed nine million and they said, "Are you sure nine is going to be enough? Why don't we set aside 14 for you just in case you need it."

Raymond Minervini II, Partner, Minervini Group and Developer of The Village at Grand Traverse Commons (Son of Ray Minervini Sr.)

The big question was, can you make a neighborhood within these functionally obsolete structures? We had this architect from France who was part of the committee to preserve Building 50. Coming from Paris, an urbanized historic part of the world, he had a different perspective about reusable structures than the typical American perspective of, "Hey, it's old. Tear it down." He was able to help us imagine that you don't have to have giant apartments. Inspired by him we have two units upstairs that are 225 square feet each. One is occupied by a guy in his 20s who works on campus, and the other is used by our construction superintendent who has his home an hour and a half way.

The low income housing component on our site was handled by arm's length co-developers. Mike and Bob Jacobson, who own Leelanau Wine Cellars, one of our state's largest wine producers based in Ann Arbor and Grand Rapids, do LIHTC [Low Income Housing Tax Credit] projects all over the state and they walked in our door one day. Honestly, it came at a pivotal time, right during the height of the Great Recession. That deal was a great cash infusion for us, and it brought a lot of housing to the site at a point where we were being thought of as a place with just a bunch of rich people living in condos. That I think really showed the community-at-large that we intended to create a neighborhood that was an entire cross-section of our community.

It is very incremental what we've been doing here, and that was what we had always imagined. We had entertained all sorts of ideas for little brewpubs and stuff. The footprint of all those buildings were really small. So can we join them together? We spent a lot of time with Earthen Ales which is now the little brewpub over here. We had all sorts of drawings to tie these two buildings together to make something big enough to be a brew house and a tap room and we finally threw up our hands and said, "What if you took this middle section of this building and we'll figure out how to fit you in there?" They were really enamored with all the qualities of that space and the timbers and the stone and the brick. And where the brew house is and Earthen Ales was there's actually a gap between two buildings and we filled that space in.

Another aspect we tried to reimagine are all these giant Kirkbride corridors. We thought, let's make them like a front porch, a shared space that can be an amenity rather than just all the space that's hard to sell. We encourage everybody to do that. "Hey, you have a bookshelf with some old books then put it out in your hallway." You know, have a reading chair with

Some of the renovated peripheral cottage buildings on the campus of the historic State Hospital complex form a localized Main Street for the community.

a lamp and you have opportunities to talk to your neighbor without necessarily having to invite them in, creating those little neutral spaces for interaction.

We had always talked about how great it would be to have an urban winery and tasting room on our campus. In the state of Michigan, there might have been one other urban winery that was a non-state production facility, and that's Left Foot Charley. He was somebody who heard about what we were doing and identified that as something he wanted to be part of. When he saw the space he said, "This is what I need." It's a 1950s building, great construction materials and robustly built. We had to do a bunch of reworking but the structure itself was great. When we get out-of-the-box entrepreneurial thinkers who see something special in what we're doing, their energy helps give the campus a flavor that we can't provide as a Disneyworld master developer.

. . .

Governor Rick Snyder came in about seven or eight years ago and was opposed to tax credits of almost any kind. He felt it was like picking winners and losers. We don't want cronyism to invade; but if a case can be made and the public interest is being served then a tax credit is a great way to say—we will give nothing until you do what you say you are going to do. Now what they've ended up doing is shifting to other types of incentive programs, and cronyism, or the potential for abuse, didn't go away; actually I think it got worse. Our local state senator has been fighting to reintroduce legislation to bring tax credits back because he understands how impactful it can be. Who knows?

Probably the bravest people were the ones that stepped forward first, without any concrete example of what was going to get built. But they were buying into the idea and they knew our company. My dad was completely invested. There was no, "Well, we'll see if this works and then we're leaving town." It was like, "No, we are here." My dad said, "It's like we sailed to the island and we burn the ship." We were totally committed.

Susan Wenzlick, Environmental Protection Agency (EPA) Brownfields Coordinator

I was a brownfield grant coordinator at the Department of Environmental Quality, now the Department of Environment, Great Lakes and Energy—Eagle. At the time I got involved, Ray had agreed to take this project on as a redevelopment. There were holes in the roof. There were trees growing through. On the floors was pigeon excrement and feathers and big chunks of peeled lead paint. I was ankle deep in lead paint walking through here—graffiti on the walls and the plaster and paint was falling off.

At that time, our grants were capped at one million dollars per project per year. I looked at this thinking, "This guy's nuts for taking this on." But, I thought if he doesn't do this the state will be stuck with it anyway because this was a state mental health facility. So either we believe in this guy and invest a million bucks or we're gonna spend the million bucks anyway.

According to Ray our initial million dollar commitment is what gave him the ability to go to the bank and get money for the roof. We cleared out the asbestos and the lead paint. In the old part of the building they have all that oak trim. Ray and his crew used a citrus-based stripper to get the lead paint off of that, and that contained the lead enough that they could just throw it away in the trash. There wasn't any special land filling necessary.

Like a lot of Great Lakes shoreline communities, there was a lot of shipping and a lot of industry here so there was a lot of contamination. Whenever people find out what I do, they say "What's contaminated?" And I say "Pretty much all of downtown is contaminated."

The main criticism our program gets is that its corporate welfare. What people don't understand is that this project would never have happened without the state's investment because the developer's bottom line doesn't work. We're leveling the playing field so that we make it appealing and attractive for developers to come to brownfield sites. Under Michigan law, the state is liable for all of this. But if it was the state's druthers we probably would have torn everything down here rather than put the money into cleaning it up and making it useful again.

The iconic spires of the historic Village at Grand Traverse Commons as seen from the barns and botanical gardens in the nearby landscape, 80 percent of which is also preserved and protected.

Russ Soyring, Director of Planning, Traverse City, MI

I remember going to the CNU [Congress for New Urbanism] conference with Ray [Minervini], and coming back in the airplane he has a little napkin from the free soda. He started to sketch out how he's going to create and redevelop this. And I said "My gosh Ray, you must be insane for even thinking that."

This town is kind of afraid of commercial intrusion in neighborhoods and expansion. But meetings about the project were led by these citizens who said, "Let's save these types of buildings," and eventually we changed the zoning code to allow for more uses to happen. The state of Michigan also had the mother of all tax breaks called a Renaissance Zone, where basically you didn't pay any state taxes or local taxes. And I think that's a really phenomenal incentive because otherwise it's easier just to do the new building thing.

Basically, Ray got the property for putting a new roof on. The state offered it to us for a dollar. But you get all the liability with it. What Ray got was maybe fifty, sixty acres; a good chunk of it in the [neighboring] Garfield Township. The township got those wooded hills as a park. Traverse City got a big wetland type of park so we put some trails through it. The idea was that about 80 percent of it was going to be green. One of my goals is to make this the most walkable community in Michigan, at least. What I really like about Ray is he wasn't going to just fill the building with people that wanted to be there. They wanted parking and he said "You're not going to have dedicated parking. Parking is going to be kind of tight. And if that's not for you then this area is not for you."

. . .

Public Act 250 was a way for the state of Michigan to declare properties blighted areas to streamline zoning to make redevelopment happen more easily. But in the beginning we made it the most difficult you could possibly have. Like, Ray wanted to change a light bulb and it took several meetings of the board to change the light bulb! Because we specified exactly the light bulb that had to go there. But I think we should never have made it so complicated. Once you have a really great planning document that the community has bought in and has basically endorsed, I think then you need to let staff make decisions.

So, we got rid of this. We created a new Planning Commission just for the Commons and we finally decided, hey you know what? Maybe we just meet quarterly. And we have to almost invent agenda items—we don't really have anything to do because everything is programmed so nicely. We just go to those three or four pages and say, "Are you putting the building in the buildable envelope? Yes. Is your parking in the right location? Yes. OK, the use is allowed."

But we are also talking about how to make things more affordable. We try to make it easier to develop. We said if you created residential development above a building you get extra building height. No parking is required for a residence above a building, outside of the downtown. These little incentives make a big difference.

Meredith Lawson, Director of Marketing, Left Foot Charley Winery at The Village at Grand Traverse Commons

Bryan [Bryan Albrich, the owner] had an inward turning left foot when he was growing up. He would take off running and forget to turn his foot out and he would trip and do a tumble like Charlie Brown, so his mom called him Left Foot Charley. To name his winery he wanted something that was fun and

With an existing natural spring in the foreground, people relax on the terrace of Left Foot Charley, a winery and hard cider maker on the grounds of the Village at Grand Traverse Commons.

inviting and not too pretentious. We don't want to scare people away from wines, because some people think wine is such a fancy drink. You don't have to be fancy, you just have to be relaxed.

. . .

We got this building in 2007 and opened the winery and tasting room. We were the first outside building to be finished, which has been cool because we've been able to watch everything come together. We do just about 20,000 cases a year in this facility, wine and hard cider as well. We work sort of like a co-op, with eighteen different growers all over the area and surrounding counties.

We acquired the barrel room in 2017—at max we can store 120 barrels. This used to be one of the old hospital's root cellars. The hospital was completely self-sustainable; the barns and farms were on the outskirts and everything was stored right here. The root cellar was built into the hill, which helps to regulate the temperature and humidity.

Old Mission Peninsula [the peninsula that juts out from Traverse City into the Bay] is where we get the majority of our grapes. The region is on the same level as some of the most famous wine areas in Germany and France. The bonus that we have is the Bay and the Lake. The bay keeps us warmer later in the year. The benefit of having vineyards all over the area is that, instead of saying that we have a fifty-acre estate and we only focus on the grape varieties suited to that spot, we can do a bunch of site tests and find the right type of soil and the right amount of sun and breeze and then plant there. Or, we'll have growers who ask us to come visit their sites and we'll go out and see that it will be best for particular varieties and they might choose to plant those, and we'll find ourselves with more Riesling or more Chardonnay.

At the start it was hard at the Grand Traverse Commons. Everyone comes past here when they're coming into town but nobody knew there was anything going on. And the roads back here were extra terrible. So this just looked like an abandoned area and nobody would come back this far. Trying to get locals to stop thinking of it as an old hospital and start thinking of it as a place where you can bike and hike and eat and drink and live was one of the biggest hurdles. Basically anyone who would stop and listen, we would tell them. It was just a lot of word-of-mouth and trying different things to try and get people here.

Half of our building is in Traverse City and half of it is in Garfield Township. The dividing line runs right through our patio. So making sure that all of the laws are the same, especially with regards to serving alcohol, wasn't easy.

Notes

1. The hospital was popular and grew. Of the 6,000 people who were residents of Traverse City in the 1920s, there were an additional 2,200 people who were patients at the State Hospital (Hains, 2001).
2. For an up to date listing of the historic structures, see: https://www.kirkbridebuildings.com/.
3. Ibid.

References

Chown, G. (2019, July 31). D. Gamble, interviewer.

The City of Traverse City (2019, February 21). The City of Traverse City Facebook page. Retrieved from www.facebook.com/cityoftc/posts/mr-raymond-minervini-sr-has-dedicated-his-life-to-restoring-and-transforming-the/2229242367298999/

Dowdall, G.W. (1996). *The Eclipse of the State Mental Hospital: Policy, Stigma and Organization.* Albany: State University of New York Press.

Egan, D. (2017). *The Death and Life of the Great Lakes.* New York: W.W. Norton & Company, Inc.

Hains, E.S. (2001). *Beauty as Therapy: Memories of the Traverse State Hospital.* Traverse City: Denali and Co.

Kirkbride Buildings (2001–2024). Retrieved from https://www.kirkbridebuildings.com/

Meza, F. (2019, August 1). D. Gamble, interviewer.

Minervini, R. (2019, August 1). D. Gamble, interviewer.

Preservation Green Lab (2011). *The Greenest Building: Quantifying the Environmental Value of Building Reuse.* National Trust for Historic Preservation.

Siciliano, P. (2021, January 24). Carol Hale, Community Caretaker. *Traverse City Record Eagle.* Retrieved from www.record-eagle.com/news/lifestyles/carol-hale-community-caretaker/article_ecf71e16-5c1b-11eb-b4ac-f3301e7557ff.html

Siple, J. (n.d.). Building Reuse is Climate Action. *Quinn Evans.* Retrieved from www.quinnevans.com/news/building-reuse-is-climate-action

Traverse City Record Eagle (2021, August 10). HALE, Carol July 2, 1942-January 5, 2021. Retrieved from https://obituaries.record-eagle.com/obituary/carol-hale-1082991988

4.2
Greensburg, KS

On the evening of May 4, 2007, a category EF5 tornado that was over 1.5 miles (2.4 km) wide decimated nearly everything in the center of Greensburg, Kansas.[1] Twelve people were killed and sixty-three were injured. With winds exceeding 200 miles per hour (321 km/hr), it traveled a total of 28 miles (45 km) and destroyed 95 percent of the buildings. That tornado was part of an outbreak that lasted three days and covered portions of Oklahoma, Colorado, Kansas and South Dakota; twenty-five *other* tornados were confirmed that day (Lawson, 2007). While most of these formed in open country, the one in Greensburg literally grew on top of the small town of about 1,400 people.

A 1915 photo captures a tornado approaching the Kiowa County town of Mullinville, Kansas, ten miles from Greensburg along Highway 54. The High Plains have experienced tornados over the centuries, but the intensity and frequency of these storms is worsening, due to impacts of global heating with climate change.
Credit: Kansas Memory, Kansas State Historical Society.

The town of Greensburg, Kansas, rebuilt green after it was leveled by a 2007 tornado.
Credit: Kiowa County Media Center.

Situated in the Great Plains of the western United States, Greensburg had experienced tornados before. The town is an 'outpost' situated in the expanse of southcentral Kansas away from the larger cities in a place so accustomed to high wind it is known as "tornado alley." However, the strength of this super cell was unlike any other; it was the strongest ever recorded. Downtown commercial buildings and surrounding homes were completely destroyed, cars were thrown hundreds of yards and trees were denuded and stripped of bark. With their houses and possessions—even their jobs—decimated, many people, naturally, left following the trauma. These departures accelerated population declines that were already happening in rural Kansas and small towns everywhere. But the remaining 700 people or so committed to rebuilding. Early on it was decided that the new Greensburg would embrace the very wind that took it down: The tiny town rebuilt for 100 percent renewable wind power, making it the city with the most Leadership in Energy and Environmental Design (LEED)-certified buildings per capita in the world (Quinn, 2013). **Today, Greensburg has not only rebuilt but it is one the greenest towns in the United States and the globe.**

The story of the town's rebuilding includes many local, state and federal resources, organizations and people across several years. Greensburg stands out among rebuilding cases, in part, because so many actors and aspects of the story defy the stereotypes of rural conservative communities, and of who participates and wields decision-making power in rebuilding. Surprisingly, the students of Kiowa County High School were early vocal leaders and steered the conversation on where to rebuild their decimated school. Local elected officials joined forces with outside advocates, and the mainstream entertainment media also arrived to usher in a green building agenda and to recruit consensus. The town was so small, it had been so devastated, and the ideas for rebuilding were so ambitious that the project captured the imaginations of politicians, designers, educators and celebrities from rural Kansas to Hollywood, California and the White House in Washington, D.C. The

Mennonite community, a motorcycling group called Patriot Riders, Kansas State college students, the Discovery Channel's Planet Green, actor Leonardo DiCaprio, and anarchists arrived to rebuild the American small town.

Standing Out

The Enhanced Fujita (EF) scale spans from zero to five, with the most damage (EF5) resulting in winds in excess of 200mph (321 km/hour). At that level, strong frame houses are lifted off their foundations and thrown considerable distances to disintegration (National Weather Service, n.d.). Trees are debarked. Unlike hurricanes, tornados are evaluated *after* the weather event actually occurs. The scale was eerily updated in 2007, just weeks before the Greensburg tornado touched down.

Upon arriving in Greensburg what one notices first is the *absence* of trees. Those that did survive the tornado are gnarly and surreal, reduced to fragmented trunks with wild offshoots. What one notices next is the block grid, a relentless street network that crisscrosses the flat Midwest terrain. There are still many empty lots. This context enables the tall vertical elements to stand out starkly against the vast horizon. In Greensburg, like in many 1800s-built railroad towns across America, the massive grain elevators and water towers, like skylines for large cities, mark the presence of the hamlet from a distance. Today, contemporary windmill farms dwarf this landscape iconography, filling the horizons *between* towns and farms, poetically reaching towards a clean energy future, visible from even farther away.

President George W. Bush on Marine One surveying the Greensburg tornado damage in the days after the storm.
Credit: Charles Dharapak/ Associated Press.

4.2 Greensburg, KS

On approach to Greensburg, Kansas, tiny grain elevators and a farmhouse are visible in the distance across a vast horizon marked by the new vertical iconography of the future, the windmill.

In an act of defiance, the region harnessed the power of the wind rather than retreating from it, and built back 100 percent green. According to the World Economic Forum, Texas, Iowa, Oklahoma and Kansas generate more than half of America's wind energy (Conte, 2022). And although wind power still only makes up 10.3 percent of America's wind energy in 2022 (Energy Information Administration, 2023), Greensburg has been able to generate more power from its wind farm than it uses. The leftover energy is put back on the grid, sold to the Kansas Power Pool. The wind farm can be seen in the distance from downtown's Main Street. Consisting of ten Suzlon 1.25-MW wind turbines with a capacity of 12.5 MW of renewable power, the array provides enough to power 4,000 modest homes. Unlike conventional fossil-power systems, the beauty of renewable energy is that it can be stored for later use, and it can be shared. Wind and solar power systems are *collaborative*.

Green in a Red State, Blue Agendas

Just a week after the tornado, Kansas Governor Kathleen Sebelius (D) declared it an opportunity to create "the greenest town in rural America" (White, 2010). This may seem too progressive for a tiny outpost in the heartland. But if some consider renewable energy and environmentalism to be left-leaning ideals, then Greensburg teaches us those ideals are also the cornerstone of this religious and farming community. The municipality upends the assumptions that politics have always divided this country and that blue and red states have nothing in common.

As a farming community, sustainability reinforced the residents' respect for the environment and the food it provided. Greensburg was also a deeply religious community. Prior to the tornado, there were seven churches for the town's 1,400 people. Going green reinforced a sense of stewardship for the earth tied to a sense of family values, faith and good citizenship (Fraga, 2012).

Rebuilding an American town from the ground up in Greensburg, Kansas. Credit: Kiowa County Media Center.

The reasons for this overlapping of seemingly divisive political agendas also had to do with the particularity of this town's rebuilding. The ambition to rebuild the town required all new construction on a virtually clean slate, so this helped the community accept something different and more visionary. The project was largely funded by outside (state and federal) sources, so this, too, helped alleviate fears that rebuilding green might be risky or a waste of local resources (White, 2010). Like many American municipalities, the town already had declining population and was strapped by fiscal austerity that deprived it of robust financial state support. But going green meant the town could afford to build back with the highest standards, since it could spend recovery money it would otherwise never have had access to for redevelopment. Years of technological advances, Big Ag and climate change had diminished the prosperity of small-town farming, so green design also promised water and energy savings (Morris, 2017). Lastly, people in the town had their children and the younger generation in mind when planning. Stephen Hardy, an urban planner on the project recalled:

> It was more about asking, "What are the things we're going to do today that are for future generations? If we have a chance to build a city for the next generation, what are the choices we can make to do that?" That framing mattered and stuck across the project. Literally "progressive" was a word that was used; but this was sixteen years ago.
> (Hardy, 2020)

Rebuilding as Fellowship

Citizen engagement and participation is generally hard to get, and robust turnout among residents can often be a challenge in urban planning endeavors. But in Greensburg following the tornado, hundreds of people came out and participated throughout the process. With far fewer resources to work with, the planning team was able to produce much more engagement through communication efforts both ad-hoc and strategic. Part of the reason for this success was residents were all personally impacted, and, there was no other way to come together as a collective and touch base. *All the buildings were gone.* So the agrarian community gathered under a huge—literal—repurposed circus tent on the outskirt of town to discuss the rebuilding plan—and to share meals, and draw near.

Planning Agency

The planning was led by the Kansas City-based design firm BNIM, who created a Long-Term Community Recovery Plan and roadmap for renewal. The town worked closely with the Federal Emergency Management Agency (FEMA), the U.S. Department of Energy (DOE) and its National Renewable Energy Laboratory (NREL), the United States Department of Agriculture (USDA), and dozens of other agencies at all levels. Sustainability quickly came to the forefront of the conversation. Even though the mayor of the town, Lonnie McCollum, resigned in the immediate wake of the tornado, the Vice Mayor John Janssen and city manager Steve Hewitt, plus environmental advocate Daniel Wallach from thirty miles away, and many others, kept momentum going and helped to form consensus.

The new town leaders, with help from BNIM, shaped municipal ordinances to ensure projects maximized the use of renewable energy resources and buildings were constructed to the highest standards of LEED. They were interested in creating contemporary, energy efficient buildings and water-conserving landscapes, among other aspects of responsible design. One ordinance mandated that all city buildings greater than 4,000 square feet (371 square meters) be rated the highest grade of sustainable design—LEED

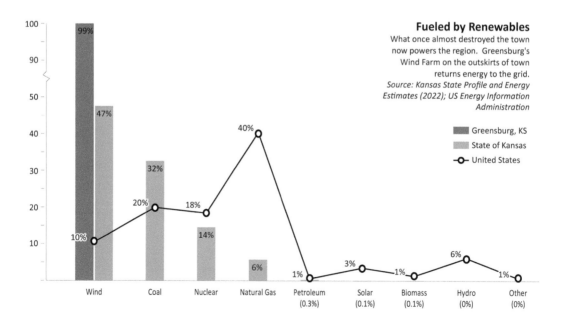

Fueled by Renewables
What once almost destroyed the town now powers the region. Greensburg's Wind Farm on the outskirts of town returns energy to the grid.
Source: *Kansas State Profile and Energy Estimates (2022); US Energy Information Administration*

Platinum.[2] Greensburg was the first town of any size in the world to do that. One of the complexities of these ambitions, however, was that disaster recovery funding and insurance protocols impractically only committed funding for communities to build back exactly what they had—not improved or updated versions. In Greensburg's case, many of the buildings destroyed were built in the last century, if not the one *prior*. In following the American public–private project, outside donations helped fill the gaps. Remarkably, nearly two decades after the disaster, the community has largely achieved their green goals, although the population continues to decline ever so slightly.

Design Capacities

The new City Hall building was designed by BNIM. It is beautifully crafted and efficient, with solar panels, maximized daylighting, rainwater harvesting and reclaimed materials: "As the first LEED Platinum City Hall project in the United States, this building incorporates a philosophy that reduces consumption at every level, reuses, recycles and cleans its waste and generates clean power for the community" (BNIM, n.d.). The streetscape outside in the center of town also uses native plantings for raingardens, with permeable surfaces. And the sidewalks along Main Street are wider than they were pre-tornado for walkability and added shade, plantings and seating.

While characteristics like this should be automatic specifications in all contemporary urban redevelopment, design strategies that challenge the norm still produce skepticism and resistance. And planning in Greensburg was not always smooth. A repeated concern in public meetings was whether or not the street would be wide enough for a combine harvester to turn around.

Public engagement in small towns generally means residents have more proportional voice in public processes, and it also means that most interactions are between leaders and residents who all know each other's name and story. Negotiations are *personal*. The Kansas town is a model of compelling design and close community collaboration. It exemplifies an urgent new set of environmental standards and cross-political, cross-resident–administration conversations all towns and cities need to be advancing.

Resources and Relationships with Agriculture

Greensburg was initially most famous not for wind but for water. In 1886, tracks from three separate railroad companies—the Atchison, Topeka & Santa Fe, the St. Louis & San Francisco, and the Chicago, Rock Island & Pacific—were being laid westward toward the small settlement. The village, which was also vying to be a county seat, was fortunate to rest above the immense Ogallala Aquifer and the town leaders knew they had a nearly unlimited supply of consistent water that could be used for the steam engines headed their way. This shallow fresh water table is one of the world's largest, and it runs below eight contiguous U.S. states. Greensburg hand dug a deep well to tap into the natural resource and the "Big Well" was born. At 109 feet (33m) deep, it provided unlimited (or so it seemed) fresh water to the town until 1925.

The population of Greensburg peaked at 2,756 in '88—1888 that is. Shortly after the arrival of the railroads, a confluence of factors led to a precipitous decline in the town's population, including poor weather, bad crops and cheaper land to the south. When the state condemned all open wells in 1933, the Big Well nevertheless remained a curiosity in Kansas. A heroic relic of a bygone era—and underground—the Big Well was one of the few structures to survive the tornado. It lives on as both a physical and cultural anchor of the town and its histories. It is also a unique tourist attraction, not unlike the limestone canal walls of Piqua, Ohio in this volume, drawing heritage tourists from around the United States and internationally.

The Big Well is now a contemporary visitor center and museum, documenting not just the growth of the town but also the 2007 tornado and subsequent rebuilding efforts. A winding stair rises two stories above the main exhibition hall to provide a 360-degree viewing platform overlooking the flat landscape. Below grade, the stair hugs the stone-lined walls that descend toward the bottom of the well. Sadly, it doesn't take long to notice how low the water table has become. Agricultural production extracts water from the aquifer at a rate *ten times* faster than it can be replenished, a frightening prospect for a portion of the country that provides a good deal of North America's nourishment. Indeed, Kansas is an arid climate, and efforts to conserve water cannot keep up with contemporary farming methods coupled with the extreme droughts of climate change. The well is a poignant reminder of the scalar imbalance that exists across the country between large "Big Ag" industries and localities who unevenly bear the brunt of their activities.

Marking the town's identity and water access, in depth and height: (left) the Greensburg water tower marks the skyline next to the rebuilt historic well building (right) with stairs that descend 109 feet (33 meters) to the aquifer that used to serve the town but is slowly drying up. The well is now a museum and visitor center.

Design Performance

Optics, both real and performative, are a major part of rebuilding American cities. Ideas need visibility, promotion and spin to generate understanding and buy-in—and to spark additional funding. A national media presence in Greensburg helped sustain interest from afar and continued raising support for the efforts long after most people outside of town had moved on.

Two catalysts that increased visibility, and thus rebuilding capacity, were the Discovery Channel's Planet Green film documentary series, and a University of Kansas (KU) School of Architecture + Design architecture studio course that would be the first to arrive and build—entirely by students and to LEED Platinum standards no less. The Planet Green series meant that multiple film crews were living nearby, or on site in trailers like the residents, for months of the rebuilding process. On the one hand, these crews brought reliability to the process of rebuilding: When the BNIM design team was back in Kansas City, they regularly communicated with the crews stationed in Greensburg to hear a report on what had happened; who said what, and other dispatches from the ground. On the other hand, the crews were always looking for a good story, and sometimes that involved artificially provoking rebuilding drama where it may not have really existed. But being filmed 24/7 put the town on its best behavior—and incentivized extra-ambitious thinking.

The University of Kansas used Greensburg as an opportunity for a graduate level design/build class. Studio 804 was an established, yearlong comprehensive educational opportunity for students entering the final year of the Master of Architecture program. The class designed and then built an art center for the community. Named 547, after the date of the tornado, the building was constructed in modules in a warehouse in Lawrence, Kansas and transported to Greensburg on flatbeds before being assembled on site. The prefabrication process involved using reclaimed lumber and the building utilized a roof-mounted photovoltaic array and a geothermal heat pump (the horizontal wind axis turbines initially installed were since removed—the wind in Greensburg proved too strong for these modest systems). There is passive cooling and water reclamation. Part art center, part community space, it was a project for residents and students to learn together and test if going green was even possible. While some in the town saw the project as too costly, where funds could have been used elsewhere, others described it as a vital community hub. For the Mayor of Greensburg, Matt Christenson, "It's where I have my weekly Dungeons and Dragons games with my fellow nerds of Greensburg." (Christenson, 2020).

Pulling Together

Today, many sites in the town center remain vacant as property owners, who lost *everything*, elected not to return to Greensburg following the tornado. The spatial effect is jarring, with many fallow and overgrown lots interspersed with contemporary infill. Some homeowners that did return to rebuild acquired adjacent properties to gain more space, instilling a more spread out, suburban character in the town than was previously the case. Almost no historic buildings remain, leaving one with an impression of incompleteness. This perception is amplified by the sparse tree canopy and the excessively wide streets (needed to accommodate combines and tractors).

It is not uncommon for municipalities to attempt to fix struggling downtowns by giving attention to the part that is in their purview: streets and sidewalks (Means, 2020). What unites the otherwise sparse downtown today is the street right-of-way. The BNIM master plan addressed the fragmented spatial pattern by reinforcing Main Street. Complete with consistent street tree planting, pedestrian-scaled lights and built-in benches, the wide walkways stretch on both sides of Main for four blocks.

The half-mile corridor is anchored on the north by the train tracks and vertical grain silo; the tall water tower and rebuilt high school bookend the axis on the south. In-between the two iconic rural anchors is The Commons, a multi-use complex that brings together the library, county history museum, university learning center and an old-fashioned soda shop on the ground floor. The second floor contains a media center. City Hall is also on Main Street next door, as well as a movie theater and variety of mixed-use buildings, including the only commercial building and one of only a few historic buildings to have survived the tornado, the S.D. Robinett Building (now listed on the National Register of Historic Places). Where Main Street crosses City Hall the brick pavers span the street, forming a raised plaza. The brick in the street is from the old city hall that was badly damaged and demolished.

The design of the corridor creates uniformity where none otherwise exists, even while the design of the new civic buildings is unabashedly contemporary.

The wide open spaces of small Greensburg, Kansas as it continues to fill back in after the devastating 2007 tornado that leveled it. Despite its size, and the vacancy that the tornado left, the town leads the world in LEED-certified buildings per capita and is a model of sustainable urbanism. People can borrow a bike to get around for free.

New commercial buildings line Main Street, anchored by one of the few historic structures to survive the tornado, the S.D. Robinett Building (at right).
Credit: Kiowa County Media Center.

The Kiowa County High School was originally planned to be rebuilt outside of Greensburg, equidistant between the two school districts and town catchments it served. But a group of engaged high school students successfully advocated to rebuild it closer to Greensburg on Main Street. Their effort revealed the power of young voices to impact rebuilding design and planning, but it also revealed the intelligence they had to understand that the school, like the other scattered buildings, needed to feel a part of the critical mass of the town. Not only was the school built back in Greensburg, it was another LEED Platinum model of responsible environmental design.

The public realm is generally the responsibility of the local government, and streets can be a town's most important civic spaces. In *Great Streets*, Allan Jacobs describes five characteristics that streets share. A great street helps make community, it is the most desirable place to be; it is physically comfortable and safe; it encourages participation; it can be remembered, leaving a long-lasting impression; and it is representative and stands as a model for others (Jacobs, 1995). Greensburg's revitalized Main Street does all this and more. It is also highly sustainable, as described above. Deep planting pits and planters are filled with native grasses. The crosswalks are raised, privileging the pedestrian, even while eighteen-wheelers cruise by. Anticipating future growth, the infrastructure investment continues one block north of Route 400/West Kansas Avenue.

Despite the elegance of the infrastructure design, community opinions vary. In a time of diminished fiscal resources in the public sector, some residents question the need for spending money on "fancy" raingardens and planters. Others recognize the importance of designing for the increased frequency of storms and rebuilding to a higher standard. What is clear is that very few municipalities have an opportunity to recast their footprint like Greensburg. Those that do have to negotiate the phasing with frustrated

Where there is still not a density of buildings along Main Street, there are trees, lights, pavers and native grasses to define and bring character to the public space and street-wall. These landscape elements double as rain-water harvesting and filtration systems in an otherwise arid climate increasingly low on water.
Credit: © Assassi/BNIM.

residents and tenants who fear construction will hurt their lifestyle, property values or business. Rebuilding here may not have restored an already diminishing population, but it provides sustenance for those who remain. And it now has one of the fastest fiber optic cable networks in the state to attract those who may still come. Above all, **Greensburg restores the balance of building and the environment, perhaps the most important metric of all for rebuilding an American town.**

Perspectives

Bob Berkebile, FAIA; Founding Principal, BNIM; and Founding Chairman, American Institute of Architects National Committee on the Environment (AIA/COTE)

In 1981, I was the architect on the Hyatt Regency in Kansas City when [two atrium sky-bridges] collapsed. I spent the longest night of my life on the rescue teams, removing more than 100 dead people, and more than 200 seriously injured people. That night my first question was, "Did I kill all these people?" But as that question was answered by failure-analysis experts and others, the real question that emerged was, *"What is the impact of my design, or my designs, on the people I intend to serve?* Am I improving their well-being, their health, their productivity, their vitality, the vitality of the neighborhood, the city, the watershed, etc.?" I found it very hard to answer those questions. And that put myself and the firm on a path. I became the founding chairman of AIA's Committee on the Environment, and I was instrumental in creating the US Green Building Council LEED rating system.

In 1993 there was the Great Mississippi Flood [across the Midwest]. I agreed to lead a national team of fifty experts to provide input for the White House and FEMA about federal policy following natural disasters. Basically we were saying, if you're going to receive federal money, you can't rebuild in harm's way. You have to build back something more resilient that brings more vitality than what was removed. That led to requests for us to help communities over the following years. Probably the most famous are New Orleans after [hurricane] Katrina, and Houston after [hurricane] Allison. And one of them became Greensburg, Kansas.

The day after the tornado, the governor called to ask if I would help rebuild Greensburg green. I told them that I was a little concerned about going to Greensburg, Kansas and saying, two days after the storm, "Hi, I'm here from the governor's office to help you." Because she was in a battle with a utility company that was planning a new coal-fired plant in their neighborhood. So I said, "Why don't we go as volunteers and listen and get to know them and if after spending some time, we think we can help and they think we could help, then I'll call you back." And so that's what put us there. Ultimately, we were hired to do the planning with FEMA and with the town to rebuild Greensburg.

. . .

In Greensburg, there was no structure left, so we had FEMA erect circus tents where people met for the first time after the storm. They celebrated the lives that were lost. And then we started this dialogue. With all the women and children present, different questions were raised than when it is usually white, middle-class men. And we then held the community to answering those questions together. It dramatically changed the outcome, including retaining the school in town, rather than locating it 15 miles (24 km) away between Greensburg and the next community. The school board had already

BNIM Architects designed Greensburg's new urban plan, but also its City Hall building, the first LEED Platinum City Hall in the United States.
Credit: © Assassi/BNIM.

purchased the land and made a deal with another district to make it a regional school. But these powerful young students stood up in a City Council meeting and challenged their school superintendent and the school board about that decision, and the board reversed it. They sold the land, and agreed to move it back to town, and rebuilt on Main Street. That obviously changed the vitality of the community as a direct result.

. . .

They also voted that they would make LEED Platinum the standard for any building that had any city or public money in it. And they took the remarkable position of *not* rebuilding their diesel-fired power plant. We were instead able to design a wind farm that generates five times the energy they consume and they can sell the excess power to the grid.

. . .

There were a lot of elderly people in Greensburg when the storm came, but essentially the population was cut in half by their not moving back. A lot of the businesses who wanted to rebuild saw that pattern and were concerned that if they build, there wouldn't be enough business to support them. They needed a way to figure that out without committing to building. So USDA helped us fund a business incubator on Main Street, which housed about fifteen businesses initially. They just paid a little bit for long enough to figure out if they can make it, and several of them now have buildings and others are still in the incubator.

. . .

John Deere [Tractor Company] had three locations in this region, one in Greensburg. They could have just not reopened. But Kelly, one of the brothers, realized right away that he needed to rebuild if he cared about this town. He built a LEED Platinum John Deere dealership and he is now also the largest wind generator distributor in North America. When Kelly sells a wind generator, he puts a hotspot transponder on it so they have perfect cell phone coverage. He's giving that community good cell service. There are a lot of those stories embedded in the town's vision of the future. In an aggregate, they helped us convince USDA and FEMA and others that that this was a time to be generous and to be creative.

. . .

When you look at the landscape on Main Street and throughout the town, you'll find that it is pretty much an indigenous landscape that doesn't require watering, and the new streetscape was designed to capture all the rainwater. Paid for by FEMA. So we, together, started trying to figure out how to live successfully with vitality and resilience, at a time when the climate is changing and agriculture is at risk in the breadbasket of America. The more we identified what we shared, the more we became allies.

Our concern was that we were going to be building back a smaller footprint than the one that was there. We wanted it to feel like a community. We

BNIM planner Stephen Hardy describes the streetscape along Main Street that his firm designed (shown here in 2010): "We put a lot of new sidewalks in. They were 20 feet wide. All the brick was harvested brick. All of these plantings, there's an underground cistern system that waters them, so they're all naturally rainwater fed. You can see the little inlets that are feeding them too. So the whole Main Street is an off-the-grid, irrigated system" (Hardy, 2020). Credit: Kiowa County Media Center.

4.2 Greensburg, KS

were arguing for building along Main Street, and trying to keep it tight initially, and then allowing it to expand. That was the basic urban design move: To respond to the highway, which wasn't going to change, and to create a spine that would have most of the pedestrian traffic and people, that would feel good and comfortable, and then allow the housing and the churches to radiate from there.

For the most part, it worked, but by the time the school district caved in to move the school location back to town, a developer had purchased what had been the school site and they were already underway with some housing. So the school had to be at the far end instead of closer in to the center of gravity. But the district liked it because they had more room for the football field and other sports fields. And we were able to do more daylighting of water management there—digging up the storm system and exposing it, so they could make it part of the biology class, so we can see how storm water could be managed and make that native landscape move through the community. Make it a part of the experience between school and outdoor classrooms and sports fields.

. . .

We bought two Priuses, because we were driving so much [between Kansas City and Greensburg] and they probably had 250,000–350,000 miles on them when we finished the work. It was a labor of love. You're pretty much married to a community that's building back from scratch. It's actually one of the reasons we don't still do that work even though we got a lot of calls afterward. I could do it because my family was grown, but for my young colleagues it was really tough to be away from their families so much.

. . .

The TV crews from the Discovery Channel wanted to film the rebuild of Greensburg and were thinking it would take something like four to six months. These are the people that film these home makeover projects, and they just thought Greensburg was a large-scale makeover. But to their credit they had film crews on site 24/7 for a year, and more than half the time for the second year. We were here in Kansas City and the film crew would call me and explain all the drama. They became our advocates because they lived it.

. . .

Our congressman Emanuel Cleaver asked me to take a busload of elected officials from Kansas City to Greensburg to see the progress. We educated them on the way down with the film from the Discovery Channel. They met the community. They had lunch. And on the drive back, one of them said, "It's too bad every community can't have a disaster." Of course I said, "Oh my God, I can't believe you said that." **But my belief is every community does have a disaster. A man-made disaster. And it's important that we raise that question without waiting for a natural disaster to reveal how broken we have become.**

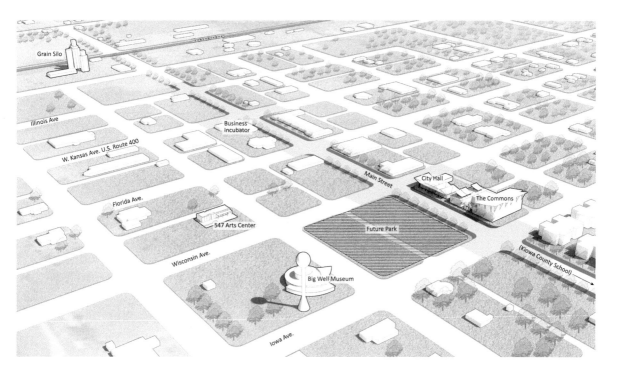

Matt Christenson, Mayor of Greensburg (since 2019)

There's very little of the pre-tornado city that remains. About two blocks on the east side of town was spared, and about half a block on the west side. Everything in the middle—just gone. That presented its own unique challenges and opportunities, in that, when you're trying to dream big, starting from that clean slate gives you a lot more freedom. But that's also just all that much more work that needs to be done—all the infrastructure has to be built from scratch. Greensburg's infrastructure was decaying at the time of the tornado. Now we have a fairly new hospital, a new school, new roads, new electrical distribution—all the core infrastructure is new as of sixteen years ago.

. . .

The city committed itself to building to LEED standard for its public buildings, but as far as private businesses, private homeowners, mandating those types of building codes would not have been feasible. So our building code since the tornado is straight International Building Code (IBC). One of the features that was incorporated into almost every public building and several private ones was geothermal heating and cooling systems. Given where we sit in relation to the groundwater table, they offer really good bang for their buck for high efficiency heating and cooling. Basically just saying, "this may cost an extra $10,000 when you install it, but over thirty years it's going to save you two or three times that in energy bills"—you find a much more receptive audience to that type of campaign than "This is what you're going do 'cause we're telling you to do it."

We operate our own electric utility as a city. If you've got any kind of generation in your home and you're producing more than you're using, we buy it back at the same rate you buy it from us. That doesn't account for all of our costs in distribution and things like that, but again it's the mindset of encouraging people to try this by making it attractive to do, because at the end of the day, you can lead a horse to water but can't make it drink.

. . .

One of the biggest issues we've had since the tornado is property owners who just never came back. And they're not maintaining their properties, mowing them, doing anything. So the city has had to spend a lot of time and effort to keep those in compliance with our codes and to make sure that we don't have places for skunks and things like that to fester. But we've got a land bank, where, if you've got a property that you don't want to take care of, you just sign the deed over to the city. Typically any back taxes are forgiven. Then if someone does want to build a house, we've got properties. If you build within a year of getting that property, it's yours for free. That's been taken advantage of several times in the last few years, which I'm happy to see.

One thing which we started to work on just before COVID hit was trying to sell smaller rural communities, including Greensburg, as a viable location for remote offices for existing businesses. My sister and her husband live in San Francisco, and he's got an hour-and-a-half commute each way, every day of the week. There's a lot of merits to small-town living: a friendly atmosphere; you're not waiting in line for everything; the cost of living is much lower; air quality is great out here—if you don't mind wind.

. . .

Decision making is more complicated [in a small town]. It's less political, more personal. If we're saying, "Sorry, no hair salons because of COVID" it's "Sorry, Debbie, you can't open your business." It's not an abstract decision for two million people. I'm making decisions for this friend and neighbor that I've known for thirty years.

Our process relied a lot on what we called *tent meetings*. We didn't have any buildings to gather in for a year or two after the tornado. So in one

One of the many extremely well-attended Greensburg community planning meetings under the big top.
Credit: FEMA/Kiowa County Media Center.

of the parks on the edge of town, we set up a big carnival tent where town meetings were held. We'd barbecue a meal, and talk about aspects of planning. What do you think about the downtown corridor? Should we have a median going down Main Street? Should we do this, that and the other? And that's how a lot of those decisions were made. People who were coming in cared. This is where they chose to rebuild their homes, their families, their lives. There was a great deal of stakeholder participation. They may not have agreed with all the decisions that were made, but they were very adamant to be a part of the process.

Alexsis Fleener, Greensburg Resident, Kiowa County High School Student and Activist After the Tornado; Community Development Director, Southwest Iowa Planning Council (at the Time of Interview); Senior Community Recovery and Mitigation Specialist, ICF

I was a freshman in high school. My grandma had been staying with us for a while and was home. I was in Salina (Kansas), getting ready to compete in the State Forensics Tournament the next day. My parents were not in town, so my grandma was the only one at our house. I was in the hotel room watching something dumb on TV when the other kids in Forensics came in and they're like, "Turn on the news, there's a big tornado coming through town." We were focused on like, "Oh my gosh so-and-so lives there, that's the southern part of town, that's where the tornado is gonna hit." It wasn't in my mind at all that—"Oh it's gonna be all of us."

My house was totally leveled. We had a big sectional couch in the basement that we never found. My grandma dove underneath a desk and was not severely injured, thank goodness.

By the morning kids still were not able to contact their family. Our return bus was stopping at different towns because that's wherever parents were going—maybe to stay with family or something. Some kids would get off. The closest we could get to Greensburg was Pratt, thirty minutes away. My mom and dad rented a house—it was me, my dads' parents, my mom's mom, and my mom, the summer after the tornado.

. . .

We never talked about, "OK, this is what we need to look for, this is what we need to find. We need to get this. This is the game plan." No, we just pulled up to the house and then it was silence. You don't even talk. You're just looking. I wasn't thinking about like "Oh I'm gonna go hang out with my friends," it was like, nope, they're at their house with their family, sifting through their stuff. Now I would consider myself a minimalist. I hardly decorate for holidays. I purge things every year. If you have so much stuff—that's just stuff to lose. Now we'd rather make memories than have stuff.

. . .

I was always interested in rural development. I got a job at Southwest Iowa Planning Council. We worked in a region of eight counties and sixty-three cities. I live here in Atlantic, which is like 6,900 people and is the biggest city in our region.

There aren't very many opportunities to do a master's degree in planning and work in rural settings in Kansas. Whereas in Iowa, every single town is a member of one of these councils of government, where you have planners and economic development people helping members in their area. We're working on water and sewer infrastructure, funding programs, downtown revitalization—we work with architects and downtown design through those programs, applying for grants through the state, and giving those small towns this whole toolbox and funding stream for their projects. That's just something that Greensburg didn't, and still doesn't, have.

. . .

My dad is a 4th generation Greensburg business owner. I lean more liberal and he's more conservative and I work in planning and he complains about the city's code of ordinances, how it's just cumbersome. I understand more the reason for the code. I help cities write them to resolve issues that they have, so I can see both sides. When I work with a small town, I also try to say "Can you enforce this? It's a good idea, it's great, but do you have the capacity to enforce this?" I think it's a balance.

. . .

It's going to be very hard for communities to just stop everything that they're doing and say "Let's reinvent ourselves" unless there is something that halts them in their tracks. I'm working with BNIM and FEMA emergency management in the city of Hamburg, which floods with the Missouri River. We're going through a re-visioning of how to build back better, how to get properties out of the flood plain, how to work with flooding but also protect the city. It's interesting because if you've flooded, there's a chance of it happening again. Whereas a tornado, the mindset is, yeah, a whole tornado went through and took away an entire town, but that's never going to happen again—so we can really think big.

In high school I always had been into social issues, having that mindset of helping others in need. Being interested in the sustainability movement was all about—how do we leave the world better than we found it? Me and a few other friends formally started a green club, and there were people who worked at GreenTown that helped facilitate our meetings. We started a recycling program and we got to go to the U.S. Green Building Conference in Boston.

Storm water strategies at the rebuilt LEED Platinum-Kiowa County High School in Greensburg, Kansas.

. . .

Dan Rockhill, JL Constant Distinguished Professor of Architecture, University of Kansas and Executive Director of **Studio 804**

Greensburg was our first LEED excursion and we were all neophytes at it back then. LEED has changed, but at the time you literally had to weigh your trash as your building waste, so we had a bathroom scale.

We were there for two months and the wind blew like crazy every day. Because of the tornado, there was absolutely no foliage on the trees, but also nothing grass-like at all. It was just this constant dust bowl and we were showering at night for half an hour to get all of that dirt out of our pores and everything else.

. . .

We're not financed by the university—Studio 804 is an independent non-for-profit corporation. I run this like a business. I would partner with community development corporations and say to them, "If you loan me $100,000 for six months, I will pay you back $108,000 after I sell the house." And they think, "We'll get the interest we would get from the bank, and we'll be able to demonstrate a successful project." But the caveat I had was that you have to let me sell that house for whatever the market will bear. So, say it was $150,000, I would make $40,000. And so I was able to then create a savings account and use that as a springboard so that now, I don't have to ask anybody for anything. I can afford to pay for the property, do the build on it, put it for sale, turn whatever profit I can, put that back in the bank, and on we go. I've been operating like that since Greensburg, which was the turning point, coincidently, for me to operate that way.

. . .

The 547 Arts Center commemorates the date of the tornado. We worked with them to put together a program for the project. We wanted to be as close as we could be to the downtown so people could walk to the project, etc., and I wanted a broad south exposure for all of the solar gain that we'd get. There were four contiguous parcels that were only fifty-foot-wide because we were on the edge of what had been residential. And we needed the last one. The students had tracked the owner down in jail in Junction City and convinced him to sign the property over to us.

We got a $100,000-dollar grant from Google, and town leaders latched onto that. [Greensburg GreenTown] said, "Oh that's for us." I'm like, "No, man. We made the inquiry with Google, we cultivated the development of the grant, and the money is for Studio 804. It doesn't have your name on it; it has nothing to do with you." For us to be able to do the project, $100,000 wasn't going to be enough. So we got a grant from the AIA for $50,000 that had to be paid back eventually, along with another Kansas housing grant that also had to be paid back.

If I waited for all the money to be in place, I probably would still be waiting. That did mean that in Greensburg, the 547 Arts Center would pay us back the deficiency. But grants were a completely foreign experience to them; these are farmers or storekeepers or whatever, so having to pursue grants was not in their wheelhouse. I knew my biggest potential failure was that they would not be able to get the funding to be able to pay us back, but I had to take the chance. So we just went ahead and did the project.

. . .

The 547 Arts Center, the first project to be built following the tornado. It is an off-grid sustainable prototype, designed and created by University of Kansas School of Architecture + Design students, led by faculty member Dan Rockhill.

You'll see Studio 804 figured in half those Discovery Channel episodes. It actually became a problem for me because they want drama, so they would create problems with the students: "So-and-so said whatever …" I finally had to get all the producers together and say, "I don't want you near this project unless you go through me." Because they were stirring it up so much and I was so determined to get done with the project. And then we finished it up and were gone. I've actually not been back. I stagger away from those projects and start another one a month later.

Stephen Hardy, Former Director of Planning, BNIM; CEO, mySidewalk

It's a very natural thing to try and build back as fast as you possibly can, but realize: you have a moment here, where you can actually change the trajectory of the community and build something that was different than what you had before. If you try to put back exactly what you had, you'll come up short. Or you can try to create a vision that's a little bit bigger.

This really positive cycle started where the city would do something that was kind of interesting, and they'd get a lot of attention and new resources. It would build on itself, and then all of a sudden, a town that nobody had ever heard of before was in the national news. There was a really nice reinforcing that would happen when they would experiment. And then NREL (the National Renewable Energy Laboratory) built them a plan for how they could harvest the wind and be net positive. So that was really good. And the rainwater harvesting: There were still cisterns literally

Map of Greensburg, Kansas before the 2007 EF5 tornado wiped out most of its trees and 95 percent of the buildings.

in people's yards. The community was used to using natural resources. They understood what it meant to use water and they knew wind was something they were always dealing with. There's something really obviously poetic about wind destroying the town, and then being something that could be a net positive for it.

. . .

A group of high school students became the emblem of the future of the city. They were entertaining, fun, smart, and they were provocative. Several of them have gone on to be city planners. And that also mattered. It was a town that was way past its last population peak in the 1960s. So the idea that there was youth engagement, and people wanting to invest in the future of the town was an enticing prospect.

4.2 Greensburg, KS

. . .

They [Discovery Channel] ended up shooting three seasons of *Greensburg*. In the beginning, three full-time camera crews were shooting eight hours a day all over the city. That had an impact. When you're being filmed all the time—you get a little used to it—but it changes the way you think about decisions. I am confident that changed the trajectory. The biggest fish out of water were these crews that had come from *Survivor* in Bali and showed up in Greensburg. They're like, what the hell did I sign up for?

. . .

It is the windiest city I have ever been to. Southwest Kansas is where the wind blows hard all the time. A twenty mile an hour wind is standard. Actually, the state of Kansas now is a net generator of electricity all because of its wind and investments in wind infrastructure, which was born out of the same governor administration that connected us with Greensburg. NREL did a whole go-to-market for how this thing could work and the city is now a net exporter of energy. Now there are lots of examples of that, but Greensburg was really early.

. . .

There were two really controversial urban design decisions. One: there's a highway and there was a proposal to build a bypass around Greensburg, with a fairly substantial argument that we should be moving commercial buildings up to the highway and abandoning Main Street. That came down to a decision that the council made, and we were strongly on one side of it and the mayor was on the other, but we came out on top in the vote. The other one was related to consolidation of the high school with Bucklin, KS. They said, "We'll put the high school halfway in between Greensburg and Bucklin. They won't

The future of the American small town is based on renewable, energy.

have to drive all the way to Greensburg, and we won't have to drive all the way to Bucklin." We were talking about how Main Street is where everybody goes to meet, it's the center of the community, the high school football games are what everybody comes to, and that's when the street is most alive. If you move your high school out of town, you're going to lose all of that, and all the businesses here will suffer. We ended up putting the high school as the southern anchor on Main Street and I think that was a really important decision.

We talked about a narrower Main Street. There's a car every seven minutes on this road, right? But there were two design requirements for the width of Main Street. It needed to be wide enough to have the homecoming parade and to accommodate the wheat harvesting thing—the thrasher—going to the grain elevator. Because they want their machinery to be driving down Main Street. And so that's part of why it stayed as wide as it did.

Notes

1. A category EF5 tornado stands for Enhanced Fujita scale. It was developed by T. Theodore Fujita in 1973 and updated (enhanced) just weeks before the tornado hit Greensburg in 2007. The scale assigns a ranking, from EF0 (65 miles per hour) to EF5 (over 200 mph) to tornados based on damage. Unlike a hurricane, a tornado's strength is rated after it hits.
2. LEED stands for Leadership in Energy and Environmental Design, a certification process by the U.S. Green Building Council.

References

BNIM (n.d.). Greensburg City Hall. Retrieved from www.bnim.com/project/greensburg-city-hall

Christenson, M. (2020, May 29). D. Gamble, interviewer.

Conte, N. (2022, April 26). Which US State Generates the Most Wind Power? There's a Clear Winner. *World Economic Forum*. Retrieved from www.weforum.org/agenda/2022/04/us-wind-electricity-generation-renewable-energy/

Fraga, R. (2012). *The Greening of Oz*. Shelbyville, KY: Wasteland Press.

Hardy, S. (2020, November 24). D. Gamble, interviewer.

Jacobs, A. (1995). *Great Streets*. Cambridge, MA: MIT Press.

Lawson, R. (2007, May 9). Barton and Ellsworth County Tornado. *NOAA*. Retrieved from www.crh.noaa.gov/ict/scripts/viewstory.php?STORY_NUMBER=2007050509

Means, M. (2020). *Main Street's Comeback*. Silver Spring, MD: Hammondwood Press.

Morris, F. (2017, May 8). 10 Years After Devastating Tornado, Greensburg, Kansas, Struggles To Move Forward. *NPR*. Retrieved from www.kcur.org/news/2017-05-08/10-years-after-devastating-tornado-greensburg-kansas-struggles-to-move-forward

National Weather Service (n.d.). The Enhanced Fujita Scale (EF Scale). Retrieved from www.weather.gov/oun/efscale

Quinn, P. (2013, April 13). After Devastating Tornado, Town is Reborn "Green". *USA Today*. Retrieved from www.usatoday.com/story/news/greenhouse/2013/04/13/greensburg-kansas/2078901/

U.S. Energy Information Administration (2023, October 20). What Is U.S. Electricity Generation by Energy Source? Retrieved from www.eia.gov/tools/faqs/faq.php?id=427&t=3

White, S.S. (2010). Out of the Rubble and Towards a Sustainable Future. *Sustainability*, 2302–2319.

Index

Note: Page numbers in *italics* refer to photographs and illustrations.

accessory dwelling units (ADUs) 159
adaptive reuse: Jamestown, NY 56–7, 66–7; North Adams, MA 26–7; San Ysidro, CA 155–6; Traverse City, MI 17, 208–9, 212–13
affordable housing 3–4; Jamestown, NY 65; Opa-locka, FL 17, 178, *185*, 186, 196–7, 200; San Ysidro, CA 152, 158–9, 160, 164–6, 169–71, *172*; Traverse City, MI 215
African American art 181–2
agriculture: Big Ag 8–9, 103, 106–7, 233, 236; Caldwell, ID 102–3, 106–7, 115–16; Greenburg, KS 232–3, 235–6; Traverse City, MI 205, 215
airports: Caldwell, ID 111; Opa-locka, FL 177, 183, 186, 191; Traverse City, MI 7, 213, 217
Alamilla, B. and Alamilla, S. 107, 118–20
Ali Baba Avenue, Opa-locka, FL 180, *181*, 186
Amazon Inc. 183, 186, 191
arts and culture: Bentonville, AR 82–3, 92; Opa-locka, FL 179, 181–2, *198*; San Ysidro, CA 155–6, 167–8; see also North Adams, MA and MASS MoCA

Ball, L. 52, 61, 68, 69
bank loans 220–1
Barnes, G. 179, *180*, 182–3, 189
Barrett, Mayor J. 44–5
Belcher, B. 35
Bentonville, AR 14, 77, *78*; art culture 82–3, 92; bike trails 84–6; global town and private philanthropy 87; Main Street 83–4; perspectives 88–96; public transit 86–7; Walmart Inc. and corporate footprint 79–80; Walton Family Foundation (WFF) 80–2, 83–4, 93, 95–6
Berger, N. 28, 40
Berkebile, B. 241–4

bicycling 10–1; trails, Bentonville, AR 84–6
Big Ag 8–9, 103, 106–7, 233, 236
Big Bling, North Adams, MA 40, *41*
Big Well, Greenberg, KS 235–6
Blackwell, M. 91–3
Bly, J. 139–41
BNIM 234, 235, 237, 238, *242*, *243*
Bonner, J. 179
Booker, K. 141–4
brownfield grants 223
Bruner, S. 41–2
Bruner/Cott 32
Bruns, R. 153
Business Improvement District, Caldwell, ID 114–15

Caldwell, ID 14–5, 101–3; "daylighting"/Indian Creek 103–6, *109*, 113–14; downtown and industrial parks 111, 121–2; flexibility 110–11; food manufacturing 106–7; Mexican food and restaurant 106–7, 118–20; multiethnic population 107–8; perspectives 112–24; satellite connections 108
Campoli, J. and MacLean, A. 4
canals: Piqua, OH 128; see also Lock Nine Park
carbon capture 212–13
Carruthers, Mayor J. 207
Casa Familiar see San Ysidro, CA and Casa Familiar
Cava, D.L. 199–200
Cava, Mayor D. L. 199–200
Chown, G. 213, 215–16
Christenson, Mayor M. 237, 245–7
Ciprazo, M. 173
climate change/sustainability 4, 8–9; building design 94, 173–4; building reuse is carbon capture 212–13; corporate footprint 80; see also flooding; Greensburg, KS
clusters see constellations

254

Collignon, K. 35–6
Community Development Block Grants (CDBG) 52, 54–5
Community Development Corporations (CDCs) 8, 152, 249; see also Opa-locka, FL and Community Development Corporation (CDC); San Ysidro, CA and Casa Familiar
Community Land Trust (CLT), San Ysidro, CA 156, 160, 162, 165, 166
Community Redevelopment Area (CRA), Opa-locka, FL 199–200
community/public engagement 157–8, 183, 233–4, 235; see also public programming
connectivity/interconnected spaces 28, 33, 156
constellations 13–4, 21–97; definition 12
cooling systems, geothermal heating and 237, 245–6
corporations: Amazon Inc. 183, 186, 191; Big Ag 8–9, 103, 106–7, 233, 236; Google 249; Walmart Inc. and corporate footprint 79–80; see also Community Development Corporations (CDCs); philanthropy; redevelopment corporations
COVID-19 pandemic 2, 4, 28, 213
Cruz, T. 157–8, 162, 168–73
Crystal Bridges Museum of American Art 82–3, 92
Cuestas, L. 166–8
Curtis, C. 173–4

Davis, C. 184, 190–2
"daylighting": canal and river, Piqua, OH 128, 132–3; Indian Creek, Caldwell, ID 103–6, *109*, 113–14
Department of Transport (DOT) 55, 65, 104–5
disinvestment/disenfranchisement 8
Downtown Revitalization Initiative (DRI) 58, 61
Dukakis, M. 34–5, 44

economic development: shifting markets, North Adams, MA 23–4; strategies, Piqua, OH 135–7; see also airports; arts and culture; tourism
Enterprise Community Partners: Rose Fellowship 153–5, 161, 162
Enterprise Zones, Opa-locka, FL 186, 191
Environmental Protection Agency (EPA) 104–5, 223

Everett, J. 120
Extreme Model Railroad and Contemporary Architecture Museum, North Adams, MA 27

facade renovation, Jamestown, NY 56, 57, *71*
Farrington, A. 45–6
Fayde, R. 198–9
Federal agencies: DOT 55, 65, 104–5; EPA 104–5, 223; FEMA 122–3, 234, 241, 242, 243
Federal funding see tax credits, *specific grants*
Ferguson, MO 8
Fitzpatrick, N. 37
Fixer Upper (TV) 10
Fleener, A. 247–8
flooding: Florida 186, 192; Ohio 132, 139–41
Flores, D. 154–5, 161, 162–6
Florida, R. 181
Forman, F. 157–8, 168–73
Fry, V. 117–18
Fultz, S. 121–2

Gaines, J. and C. 10
Gamble, D. 142
Gebbie Foundation 50, 57, 63, 65, 70–2
gentrification 181, 186, 196
geothermal heating and cooling systems 237, 245–6
global border conditions see San Ysidro, CA
global and local relationships 30–1, 87
global and national influences 8–10
Goody Clancy 50, 57, 62–5
Google 249
Great Miami River, Piqua, OH 128–9, 131–2, 141, 142
Greensberg, KS 17–8; design capacities 235; design performance 236–7; Discovery Channel 10, 18, 237, 244, 250; green policy and political agendas 232–3; "green" rebuild 9; perspectives 241–53; planning agency 234–5; public participation/engagement 233–4, 235; pulling together 237–40; railroads and Big Well 235–6; tornado 5, 8, 229–31; wind power 230, 231–2, 253
Gunderson, J. 68–70

Hahle, M. 123
Hale, Mayor C. 206
Hardy, S. 233, *243*, 250–3
Heller, K. 88–90
Heritage Park, North Adams, MA 26, 27
historic preservation 56
historic tax credits 54–5, 56, 209
Home Town/Home Town Takeover (HGTV) 10
Hood, W. 180, 187, 192–4
hotels/motels, North Adams, MA 27, 30, 36, 43–4
housing, affordable *see* affordable housing
Housing and Urban Development (HUD) grants 178
Huff, G. 129, 138–9

inflation and affordability crisis 3–4
International Building Code (IBC) 245–6

Jackson, R.A.M. 52
Jacobs, A. 239
Jamestown, NY 14, 49–50; building stock and revitalization 55–7; core planning 57–8; National Comedy Center 52–5, *59*, 61, 68–70; perspectives 61–74; railroad and station 50–2; region relations 58–60
Joy, V. 66–7

Kirkbride, T.S. 207
Krens, T. 34, 41
Krzyston, C. 216–17

Land Conservancy (GTRLC) charity, Traverse City, MI 211–12, 215–16
Lawson, M. 225–6
LEED (Leadership in Energy and Environmental Design) 115, 161–2; Greensburg, KS 230, 234–5, 237, 239, 242, *248*, 249
Leingang, M. 129
Living Rooms at the Border and Casitas project, San Ysidro, CA 152, 158–9, 164–6
Lock Nine Park, Piqua, OH *130*, 131–2, 133, 134–6, 142–4
Logan, W. 194–7
low-income housing tax credits 164, 221
Luoni, S. 91

McGee, T. 35
MacKeith, P. 93–6
Main Streets: aura of 5–6; Bentonville, AR 83–4; Greensburg, KS 238, 239, *240*, 242, 243–4, 252–3
Malone, J. 122–3
map of towns and cities *1*
Massachusetts Museum of Contemporary Art *see* North Adams, MA and MASS MoCA
Mayors: Barrett (North Adams, MA) 44–5; Carruthers (Traverse City, MI) 207; Cava (Opa-locka, FL) 199–200; Christenson (Greensburg, KS) 237, 245–7; Hale (Traverse City, MI) 206; Nancolas (Caldwell, ID) 104, 105, 112–15; Teresi (Jamestown, NY) 62–3, 65
mental health hospital, Traverse City, MI 205–7
methodology 18–9
metroposts 16–7, 149–201; definition 12
Mexican food and restaurant, Caldwell, ID 106–7, 118–20
Mexican-American laborers, Caldwell, ID 106
Meza, F. 212, 218
mills: Jamestown, NY 50; North Adams, MA 23, 24, 25, *32*, *39*, 42
Minervini, R. 208, 211, 215, 218–21, 223, 224, 225
Minervini II, R. 221–3
Minervini Group 206–7
Minkel, K. 88–90
mobility *see* bicycling, public transit, railroads
Mohawk Trail (Route 2) 30, 43
Moorish designs, Opa-locka, FL 16, *179*, 180, 182, *184*, 189, *190*, 193–4
Morris, E. 73–4
multiethnic populations: Caldwell, ID 107–8; Opa-locka, FL 193–4
multinational industries *see* corporations
Mumford, L. 54
Museum of Glass, Corning 58

Nancolas, Mayor G. 104, 105, 112–15
Napier, E. and B. 10
National Baseball Hall of Fame and Museum, Cooperstown 58
National Comedy Center, Jamestown, NY 52–5, *59*, 61, 68–70

National Endowment for the Arts 162, 178
national media and TV 10; Greenberg, KS on Discovery Channel 10, 18, 237, 244, 250
National Register of Historic Districts, Jamestown, NY 56
National Renewable Energy Laboratory (NREL) 250–1, 252
national themes and impacts 2–7
neoliberalism 170
New Market Tax Credits 67, 68, 167, 171
Ney, L. 70–2
Nixon, A. 72
North Adams, MA and MASS MoCA 13; Berkshire constellation/arts and education project 28–30; critical mass 25–8, 30; developing arts 31; dimensions of rebuilding/early renewal strategies 24–5; economic shifts 23–4; global local relationships 30–1; perspectives 34–47; subtraction strategy 31–3
nostalgia 2

Oda, J. 142
Opa-locka, FL and Community Development Corporation (CDC) 16–7, 177; designing process 182–3; development climate 186–8; perspectives 189–200; reclaiming "creative class" 181–2; scales of shared power/planning 177–80; stakeholders 183–6
Oppenheimer, S. 108, 122–3
outposts 17–8, 203–53; definition 12

philanthropy: Gebbie Foundation 50, 57, 63, 65, 70–2; Wadsworth 35–7, 38–9; Walton Family Foundation (WFF) 80–2, 83–4, 87, 88–90, 93, 95–6
Piqua, OH 15, 127–9; canal and river 128, 132–3; community 129–30; investment 133–4; perspectives 138–46; planning initiatives 130–2; strategies 135–7
political and social crises 2
population loss 6–7
Porches Inn, North Adams, MA 27, 36
Preciado, O. 173
private foundations/finance see philanthropy, public–private partnerships

public programming 11, 106; see also community/public engagement
public transit: Bentonville, AR 86–7; San Ysidro, CA 159–60, 167–8
public–private partnerships 8, 11; Bentonville, AR 95–6; Caldwell, ID 105; Jamestown, NY 57–8, 70; North Adams, MA 30–1; Piqua, OH 136–7; Traverse City, MI 205–7

railroads: Caldwell, ID 108, *109*; Greensburg, KS 235, 236; Jamestown, NY 50–2, 60; North Adams, MA 26–7; Opa-locka, FL 186–7, 196, *197*, 198–9; San Ysidro, CA 159–60
redevelopment corporations 52, 215
Raine, A. 35
remote working, rise of 2, 4, 213
Renaissance Zone, Michigan 209, 224
RFP (request for proposal) grant 194–5
Rice, B. 66–7
Rockhill, D. 249–50
Rose Fellowship: Enterprise Community Partners 153–5, 161, 162

San Ysidro, CA and Casa Familiar 16, 151–2; community builders (CDCs) 155; contemporary contextualism 157–9; leading by design 153–5; local arts and culture/culture corridor 155–6, 167–8; negotiating scales 152–3; perspectives 161–74; transit orientations 159–60, 167–8
satellites 14–5, 99–147; definition 12
scale matters 7–8
Scenic Byway, Mass-NY 30
Schmiesing, C. 137
Sebelius, K. 232
shared trends 7–11
Smith-Sigman, K. 115–17
social and political crises 2
Soyring, R. 224–5
spatial structures 11–8
Spillane, D. 62–5
sprawl 4; Caldwell, ID 108, 115
Stayner, C. 179
Steves, J. 73–4
Stirratt, J. 30, 43
sustainability see climate change/sustainability
Svenson, B. 30, 43–4
Swenson, K. 161–2

tax credits: historic 54–5, 56, 209; low-income housing 164, 221; new market 67, 68, 167, 171
tax free Renaissance Zone, Michigan 209, 224
Teresi, Mayor S. 62–3, 65
Teresi, S. 62–3, 65
Thompson, J. 37–40, 42
TIGER funds 90
tourism: Caldwell, ID 115–17; critique of 53; Greensburg, KS 236; North Adams, MA 26–7, 28, 30, 37–40, 43–4; see also Jamestown, NY
Town Common, North Adams, MA 26, 27
trauma and collective support 8
Traverse City, MI 17, 205–7, 214; building reuse is carbon capture 212–13; Land Conservancy (GTRLC) charity 211–12, 215–16; landscape connections as therapy 207–8; locally isolated but nationally connected 213; mixed-use Village and "mercato" 210–11; perspectives 215–26; starting small 208–9, 210
Turrell, J. 32

typologies see constellations, metroposts, outposts, satellites

University of Kansas (KU) School of Architecture + Design 237, 249–50
urban renewal 24–5, 40, 56, 58

Wadsworth, J. 31, 35–7, 38–40
Walmart Inc. and corporate footprint, Bentonville, AR 79–80
Walton, A. 92
Walton Family Foundation (WFF) 80–2, 83–4, 93, 95–6
Water Industry Act (1991) 104
Watson, S. and Watson, S. 144–6
Weld, B. 35, 44
Wenzlick, S. 223
Wilco 30, 36, 38, 43
Williams College, North Adams, MA 28–9, 34, 38
Williamstown, North Adams, MA 28–30, 38
wind power, Greenberg, KS 230, 231–2, 253

Yawn, E. 45–6

Zemsky, H. 61–2